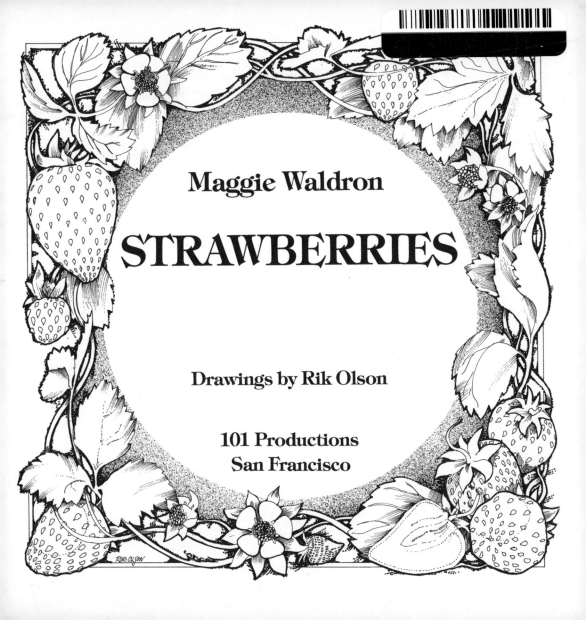

Maggie Waldron

STRAWBERRIES

Drawings by Rik Olson

101 Productions
San Francisco

FOR SARA

Who, at the age of two, knew that fresh, warm San Francisco sourdough (double-baked crusty, the kind you get at the wharf), slathered with butter and piled with sugary strawberries, was the Most.

Copyright 1977 Maggie Waldron
Drawings copyright 1977 Rik Olson

All rights reserved. No part of this book may be reproduced in any form without the permission of 101 Productions.

Distributed to the book trade in the United States by Charles Scribner's Sons, New York, and in Canada by Van Nostrand Reinhold Ltd., Toronto

Published by 101 Productions
834 Mission Street
San Francisco, California 94103

Library of Congress Cataloging in Publication Data

Waldron, Maggie
   Strawberries.

   (Edible garden series)
   Includes index.
   1. Cookery (Strawberries)   I. Title. II. Series.
TX813.S9W34   641.6'4'75   77-2126
ISBN 0-89286-112-6

contents

Strawberries Stem from the Rose Family 5
Strawberries Ripe 6
Strawberry Arithmetic 7
Your Own Homegrown 8
Strawberries Preserved 12
Les Crèmes des Crèmes 26
Breakfast Any Time of the Day 38
Sweet and Sharp Salads and Compotes 52
Refreshingly Frozen Drinks and Desserts 62
Dazzling Desserts 76
Index to Recipes 94

STRAWBERRIES

Curlylocks, Curlylocks,
Wilt thou be mine?
Thou shalt not wash dishes
Nor yet feed the swine,
But sit on a cushion
And sew a fine seam,
And feed upon strawberries,
Sugar and cream.

—*Anonymous Nursery Rhyme*

strawberries stem from the rose family

Sniff deeply. Strawberries might well be the real roses of Picardy. For it was during the reign of Louis XIV in France that the marriage of two New World varieties took place—a wild meadow berry from Virginia with a large, bright Chilean—becoming the pollen parents of our flamboyant American beauty of today.

Wild strawberries grow in many climates on many continents. The French *fraises des bois,* "strawberries of the woods," were cultivated as far back as the 15th century with glorious acclaim by the likes of Ovid and Pliny. The early colonists in North America discovered them in wild abundance, a true wonder fruit. The Indians crushed them over warm cornbread and our Pilgrim Fathers sliced them over Indian pudding (ideas I find so enticing you'll find recipes herein).

Even though American horticulturists and others made giant strides in the breeding of the berry, it remained pretty much a home-grown, limited-season crop until the University of California Agricultural Station introduced exciting new varieties. Since then, there has been no stopping this big, sun-blessed Californian. The strawberry yield per acre has almost tripled. Experimentation continues to produce bigger and better berries with longer growing periods. They are now being shipped all over the world.

Since the English very likely had something to do with taming and naming this berry, it seems fitting that an Englishman have the final word: "Doubtless God could have made a better berry, but doubtless God never did."

strawberries ripe

"Streowbery ripe!" cried the early London street vendors. A far cry from today when strawberries are ripe, or air-freight ripe, from February through November (to say nothing of fresh-frozen ripe all year long). However, May is strawberry month, the time when production is heaviest, the time when irresistible berry patches appear in every market.

Select berries that are bright red and firm, for they do not ripen appreciably after being picked. Make sure that the caps are fresh and green. Strawberries are best kept on a shallow tray, loosely covered, in the refrigerator. Plan to use them within a day or two.

Do not wash berries or remove caps until just before using. Otherwise you will wash away their natural protective coating. To wash, rinse whole berries with their caps on in a gentle spray of cool water. Drain well. Unless the recipe specifies "with stems," remove caps by twisting off or use a strawberry huller or the end of a sharp paring knife. *All the recipes in this book require that the strawberries be washed and capped just before using unless otherwise directed.*

Mind the condition of your berries before you use them. Musty or green berries won't do at all. Soft berries do very well in ices, syrups, sauces, and so on. Ordinary berries make extraordinary salads and desserts. But the big, breathtaking berries are simply best—undressed.

IMPORTANT NOTES ON USING THESE RECIPES
- *Strawberries should always be washed just before using.*
- *Always remove caps unless recipe specifically directs otherwise.*
- *One "basket" of strawberries refers to the market package and is equivalent to about 3-1/2 cups of whole berries.*
- *Remember to set aside a few berries for garnish.*

strawberry arithmetic

FRESH WHOLE STRAWBERRIES
1 basket strawberries = about 3-1/4 cups whole berries
1 basket strawberries = about 2-1/4 cups sliced berries
1 basket strawberries = about 1-2/3 cups puréed berries
1 basket strawberries = from 12 very large to 36 small stemmed berries
1 cup whole strawberries = about 4 ounces
1 tray or flat of strawberries = 12 baskets

FROZEN WHOLE STRAWBERRIES
20-ounce bag frozen strawberries = about 4 cups whole berries
20-ounce bag frozen strawberries = about 2-1/2 cups sliced berries
20-ounce bag frozen strawberries = about 2-1/4 cups puréed berries
10-ounce package frozen sliced strawberries = 1-1/4 cups berries in syrup

STRAWBERRY NUTRITION
Strawberries are an excellent source of Vitamin C, with 1 cup supplying about 150 percent of the U.S. Recommended Daily Allowance for the average adult. Fresh or frozen berries also provide iron and other minerals—1 cup of whole strawberries provides about 8 percent of the U.S. RDA for iron and 2 grams of natural fiber.

BEST NEWS YET:
1 cup of whole, unsweetened strawberries has only about 60 calories.

your own homegrown

Just how straw became so closely associated with this berry is subject to conjecture. But in early England, strawberries were ripe at the time the hay was mown—and to this day the berries are bedded down with a covering of straw in winter; in spring, the straw is spread as a mulch between the rows to promote even ripening.

Where to plant If you have the space, you'll want to plant as many berries as you can, for they are one of the best fruits to freeze and preserve. You can expect about a pint of fruit from each plant in a season. Since they are decorative as well as delectable, you can grow them as a border around a flower or vegetable garden, in a raised bed, in pots, or in an old-time "strawberry barrel" with 2-inch round holes in the sides for inserting the plants as well as the usual top opening.

What varieties are best for you The everbearing varieties will reward you with two harvests—the most bountiful in early summer and another one in early fall, but you can stretch your own season by planting several varieties. Catalogue descriptions will help you with your choice of early-to-midseason-to-late varieties. Check with a reliable nursery or call your Farm Bureau for good local recommendations, since each region has unique climate and soil conditions. Most important, buy disease-free plants from a reputable nursery. There are some very good everbearer varieties coming on the market at this time—the Quinalt is the most disease-free everbearer tested. The varieties that have good track records throughout most of this country are the Ozark Beauty and the new Fort Laramie.

When to plant In the South and other mild winter climates, late summer or fall is best for bountiful spring harvests. In colder climates, early spring is the time to plant.

Sun and soil Strawberries generally grow best in full sun, in fertile, well-drained, slightly acid loam soils. Add compost, rotted leaves, peat moss, chopped seaweed, or any other good organic materials available to you. Mix in a complete fertilizer such as 5-10-5 at the rate of 4 pounds to 100 square feet. Sandy soils will require a second application of fertilizer in midsummer. No fertilizer beyond the initial one at planting time should be given in future springs—too much fertilizing will cause soft berries and contribute to rot.

Spacing There are different systems and differing views around the country—again, your County Agent can most likely give you detailed instructions.

Single rows of strawberries: Set plants from 12 to 18 inches apart.

Double, triple, or multiple rows: Set the plants 12 to 18 inches apart in rows; allow 18 to 24 inches between rows according to the space available. Bear in mind that you're going to have to pick the berries in this space.

Planting Your young plants should be kept moist while planting. Do not crowd the roots in the planting hole; allow enough room for the roots to point down into the earth. Keep the center of the crown level with the soil surface with the top of the crown exposed and all the roots covered. Tamp around each plant so roots and soil are in firm contact. Water well at once.

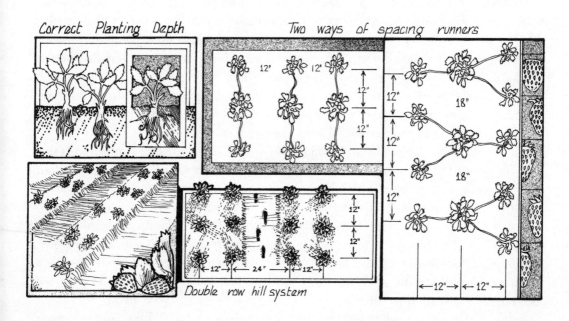

Care of ripening plants Strawberries require frequent watering and should never be allowed to dry out completely. Leaves may be sprayed with a water-soluble fertilizer designed for that purpose.

Always mulch the plants, no matter how or where you grow them. Mulching keeps the soil moist, discourages weeds, and lifts ripening berries off soil. Severe climates require heavy mulching.

As plants mature, shoots called runners will fan out from the plants. The number of runners that are left attached to root and form new plants will determine the size and number of berries in the crop. Some gardeners prefer to sever and discard runners as soon as they appear to insure large but fewer berries. Others allow some or all of the runners to root, thereby forming new plants and a crop of smaller, but more numerous fruits. You may also sever the runners with the immature plants attached and root them in a new bed or container.

Cut off the flowers that appear the first summer after planting, but allow the later flowers of everbearing varieties to form fruits for a fall harvest. As fruit begins to appear, you may want to cover plants with a wire or coarse jute mesh to protect from birds. Each year your plants will bear fewer, smaller berries; replace plants after 3 years.

Harvesting Constantly check plants for ripe fruit as it will rot quickly if left on the vine. Pick your berries as they become fully red/ripe, for they won't ripen appreciably after picking. Chill them fast. Never wash them or remove caps until just before using; washing removes their natural protective coating.

Pests and diseases Check undersides of ailing leaves for signs of aphids or other pests; a spray of soapy water should discourage any invasion. As further preventatives, choose a variety that is disease resistant and keep your berry garden well tended. If invaders do appear, act quickly: Use nicotine spray or other organic means of control. Avoid spraying ripening fruit.

strawberries preserved

This section is going to deal with preserving the flavor of this beautiful berry. And not just for jams and jellies. Strawberries are one of the best fruits for freezing and, certainly, the easiest. You can sun-cook them into preserves, dehydrate them into leather candy, and *Rumtopf* them with other fruits into year-round indulgences—all part of the preserving game.

PREPARATION OF YOUR FRUIT
However you plan to use the strawberries, sort through them and separate the soft ones from the firm, fully ripe berries. Discard any musty or spoiled berries. Get to your preserving pronto, using the softer berries in syrup, purée, leather, jelly.

MAKING JAMS AND JELLIES AND OTHER STRAWBERRY PRESERVES
Equipment you'll need
large 6- to 8-quart enamel or stainless-steel kettle with a flat bottom,
   no cracks or chips inside
ladle, large metal spoon, slotted spoon or skimmer
jelly bag, cheesecloth or muslin
colander
large bowls
wide-mouth jar-filling funnel
jars, jelly glasses, lids, paraffin
timer

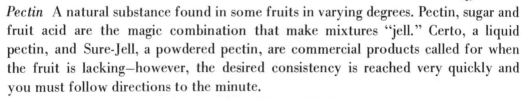

Pectin A natural substance found in some fruits in varying degrees. Pectin, sugar and fruit acid are the magic combination that make mixtures "jell." Certo, a liquid pectin, and Sure-Jell, a powdered pectin, are commercial products called for when the fruit is lacking—however, the desired consistency is reached very quickly and you must follow directions to the minute.

The jellying point When making jelly without added pectin it is absolutely essential to cook the boiling syrup to exactly the right point. Thermometers are helpful (220° to 222°F is the jelly stage), but the sheet test should also be used, as jellying points aren't always the same. To make the sheet test, dip a large metal spoon into the boiling syrup. Tilt the spoon until the syrup runs from the side. When the jellying point has been reached, liquid will not flow in a stream but will divide into two distinct drops that run together and will "sheet" from the spoon.

How to sterilize, fill and seal jars Use glass jars that can be sealed airtight—straight-sided or regular canning jars. Wash jars in hot, soapy water and rinse. Sterilize by placing in hot water and bringing to a boil. Let stand in boiling water at least 10 minutes. Keep the jars in hot water until ready to use.

If using screw-top jars, sterilize the lids according to manufacturer's directions. A thin paraffin seal has more charm and allows the shimmering jelly to be turned out intact. Since paraffin is highly inflammable, it should never be melted over direct heat. You might keep a "paraffin pot" for this purpose—put the paraffin in the pot and put the pot in a pan of simmering water to melt it. Any leftover paraffin can be reheated for the next batch of jelly.

Fill hot jars with hot jelly or fruit mixture to within about 1/4 inch of tops. If using lids, put on each jar as it is filled and screw band tight. If using paraffin, cover immediately with 1/8 inch hot paraffin. To seal properly, rotate the glass so the paraffin climbs up the sides, making an airtight seal. Do not be tempted to use two layers of paraffin because it is too heavy and will not hold a seal. Place sealed jars away from drafts until cool. Label, date and store in a cool place.

JACQUES PEPIN'S FRAISES AU SOLEIL
(Strawberries in the Sun)

"This is the most delicious strawberry preserve I have ever tasted. The whole berries are actually candied in the sugar and remain almost intact through the whole procedure." Coming from Jacques (author, *House Beautiful* chef, restaurateur, connoisseur), you know these will be sensational.

3 pounds sugar	4 baskets (approximately 3 pounds) nice whole strawberries, well ripened but not bruised
3/4 cup water	

Mix the sugar and water together. You will have barely enough water to moisten the whole amount of sugar. Place on high heat and bring to a boil. Keep boiling for 5 minutes. The sugar may look whitish and crystallized, but it is all right. Add the strawberries to the syrup. Bring to a boil and cook 5 minutes. The berries should release their juices and get lumpy, and the syrup should become smooth, liquid and pink. Pour the mixture into a large roasting pan, spreading it so it is no thicker than approximately 2 inches. Place in the sun covered with a window screen to inhibit flies, bees, and the like. Take back in the house at night. It should take about 4 days of sun. The liquid will slowly evaporate and the berries will soon become swollen again with the heavy syrup. You will notice that after approximately 4 days, the syrup is very thick and will not reduce any more. Ladle at once into hot, sterilized jars and seal.
Makes about 2 pints

COUNTY FAIR STRAWBERRY PRESERVES

A great cook and my good friend, Eleanor Tomic, claims the secret in making these marvelous preserves is adding the berries a few at a time so that the syrup never stops boiling.

4 cups sugar
2/3 cup water

4 cups beautiful strawberries
1/2 cup fresh lemon juice (and a jigger of brandy, if you like)

Bring sugar and water to a boil, stirring until sugar has dissolved. Boil gently 5 minutes. Now add berries a few at a time so that the boiling never stops. Gradually add lemon juice. Boil the whole batch gently, without stirring, for 18 minutes. Skim off foam. Pour into a shallow pan (13 by 9 inches). Let stand 24 hours, shaking pan occasionally. Pour into hot, sterilized jars and seal immediately.
Makes about 2 pints

Note
- The more beautiful the berries, the more beautiful the preserves.
- Make just 1 batch at a time—it won't do to double this recipe.
- Remember that most preserves need to be stored about 2 weeks to reach the proper consistency.

QUICK COUNTY FAIR STRAWBERRY PRESERVES

2 baskets strawberries
7 cups sugar (3 pounds)
1/4 cup fresh lemon juice
3 ounces liquid pectin (half of a 6-ounce bottle)

Measure 5 cups whole, ripe berries (a packed measurement, but without crushing). Layer in broad, heavy pan with sugar. Let stand 10 minutes. Bring slowly to boil, stirring gently to keep fruit whole. Remove from heat. Cool at room temperature 4 hours. Add lemon juice. Bring mixture to full rolling boil over high heat; boil hard 2 minutes, stirring gently. Remove from heat; at once, stir in liquid pectin. Skim off foam with metal spoon and stir for 10 minutes, to prevent floating fruit. Ladle into sterilized jars. Seal at once with lids or paraffin.
Makes about 3-1/2 pints

STRAWBERRY AND RHUBARB PRESERVES

4 cups rhubarb, cut in 2-inch pieces
8 cups sugar
4 baskets strawberries
1/4 cup fresh lemon juice

Toss the rhubarb with sugar and let stand 12 hours or overnight. In a heavy kettle, bring rhubarb and sugar to a boil. Add berries gradually, keeping mixture boiling all the time. Boil mixture gently until thick, stirring occasionally. Remove from heat. Stir in lemon juice and pour into sterilized jars. Seal immediately.
Makes about 4 pints

ROSE CRYSTAL STRAWBERRY JELLY

5 baskets strawberries (about 15 cups)
7-1/2 cups sugar

1/2 cup strained fresh lemon juice
1 6-ounce bottle liquid pectin

Crush the berries. Mix with 2 cups of the sugar in a large kettle. Bring to a boil, then simmer 5 minutes. Pour mixture into jelly bag or through a strainer lined with several layers of cheesecloth. Let juices drip without squeezing to keep jelly crystal clear. Measure 3-1/2 cups of juice back into kettle. Stir in remaining sugar and lemon juice. Bring to a boil over high heat, stirring all the time. Stir in pectin and bring to a full, rolling boil. Boil hard 1 minute, stirring constantly. Remove from heat, skim off foam with metal spoon, and pour quickly into hot, sterilized jars. Seal with lids or cover at once with 1/8 inch paraffin.
Makes about 7 cups

STRAWBERRY LEMONADE JELLY

2 lemons
5 cups cold water

1 basket strawberries, crushed
4 cups sugar

Slice lemons very thin and place in bowl with water. Cover and let stand 18 hours. Pour into a stainless steel saucepan with strawberries; bring to a boil and simmer, covered, 40 minutes. Line colander with several layers of cheesecloth or use jelly bag. Pour mixture through, letting it drip undisturbed. When mixture stops dripping measure juice. If it measures more than 4 cups, boil rapidly to reduce to 4 cups. Mix with sugar in large kettle. Boil rapidly until mixture sheets off spoon. Remove from heat, skim foam and ladle into hot jelly glasses. Cover at once with 1/8 inch paraffin.
Makes about 4 8-ounce jars

HELEN McCULLY'S FAMOUS STRAWBERRY SYRUP

Helen McCully, *House Beautiful's* incomparable food editor, is also famous as a warm, witty, compassionate human being. The bottling of the berry is an old European custom—lovely for fizzy drinks, ice cream, pancakes, even oatmeal. This is Helen's way of making it.

4 cups fresh ripe strawberries   about 2 cups sugar
1 cup water

Wash, drain and hull the berries. Place in a saucepan with 1 cup water. Bring to a boil, reduce heat, and simmer *exactly 10 minutes,* no more. The fresh flavor depends on minimal cooking. Now comes the tricky part. You need to strain off all the juice through a jelly bag. If you don't have a jelly bag, use several layers of cheesecloth. You will have to devise some way to hold the fruit in the cloth above a large bowl so it can drip. A little judicious squeezing is allowed to extract the maximum amount of juice. Measure the juice into a saucepan and discard the pulp. Add 1 cup of sugar for each cup of juice. Cook over moderate heat, stirring, until the sugar has dissolved and the syrup comes to a boil. Boil *exactly* 2 minutes. Remove from heat and skim off the froth. Pour the hot syrup into hot sterilized jars, leaving 1/2-inch headspace, and seal.
Makes about 1 pint

Note The exact amount depends on the ripeness and juiciness of the fruit. This recipe can be doubled or tripled with equally good results.

A PERPETUAL POT OF FRUIT WITH RUM OR BRANDY

This is a good way to use fruit as it comes into season, adding more syrup, spirits, or fruit to keep your well from running dry.

4 cups water
4 cups sugar
1 cup fresh lemon juice

1 cup rum or brandy
assorted fresh fruits, see below

Bring water and sugar to a boil, stirring until sugar dissolves. Simmer 15 minutes. Cool. Add lemon juice and rum or brandy (or some of both). Pour into a large container with cover. Add fruit to the syrup, making sure that the fruit is submerged. Refrigerate, replenishing as needed. Keeps beautifully.

Fruit Suggestions
- Fresh whole berries. Strawberries, blackberries, raspberries, blueberries, loganberries, gooseberries.
- Cherries. Remove stems and pits.
- Plums, peaches, pears, apricots, nectarines. Cut in half, remove pits or cores, and quarter or cut into chunks.
- Grapes. Pull grapes from stems, halve, and seed; or, use Thompson seedless or Perlettes whole.
- Pineapple. Top and tail pineapple, cut away rind. Core and cut into chunks.
- Papaya. Cut in half, peel, and scoop out seeds. Slice or cut into chunks.

HOW TO KEEP YOUR FRUIT ICY-CRISP

You can keep a perpetual mélange of fruit going in this simple, perfectly proportioned syrup that does indeed keep fruit icy-crisp.

1 cup sugar
1 cup water
1/4 cup fresh lemon juice
2 teaspoons grated lemon peel
up to 1-1/2 quarts whole strawberries and other fruit

Bring sugar and water to a boil, stirring to dissolve sugar. Boil rapidly 5 minutes. Chill well. Stir in lemon juice and peel. Cut up berries and fruit, dropping pieces into syrup as you go—this amount of syrup will take about 1-1/2 quarts of fruit. Cover and chill well. Will keep a week or longer.

Frosted strawberries are beguiling around such dishes as cold roast chicken, fresh fruit salads, and sandwich plates. Simply dip the berries into frothy egg whites, then into coarse sugar, and let them dry on cake racks. Remember strawberries, too, as a garnish for all those creamy, subtle dishes that need a splash of color on the side.

FREEZING STRAWBERRIES

Equipment you'll need
a true freezer capable of maintaining a temperature of 0°F is important to the freezer life of all fruits and vegetables—up to a year for strawberries
containers or wrappers (any of the following): slope-sided, dual-purpose freezing and canning jars; plastic freezer containers with lids; plastic freezer bags; freezer-weight aluminum foil with freezer tape; plastic freezer film; or waxed or plastic-coated freezer paper
labels with room for recipe name and date

The unsweetened pack Strawberries are one of the few fruits that can be packed perfectly plain and unsweetened. Simply rinse the whole berries and drain them; cap, pack, seal, and label. The unsweetened pack requires little or no headspace.

The dry-sugar pack Gently mix whole or sliced berries with sugar (6 parts berries to 1 part sugar), turning several times until sugar has dissolved. Pack in containers with a crumbled piece of freezer paper on top to keep fruit under the juice before sealing with lid.

The syrup pack Dissolve 4-3/4 cups sugar in 4 cups hot water, substituting fresh lemon juice, wine or fruit-flavored spirits to taste for part of the water. Cool syrup. Pack fruit in rigid containers and cover with syrup, allowing 1/2-inch headspace for pints, 1 inch for quarts. Crumble a piece of freezer paper to hold fruit under syrup before freezing.

BRIGHT STRAWBERRY SAUCE

Have you ever had a hot strawberry sundae?

2 cups water	2 tablespoons cornstarch
2 cups sugar	1/4 cup fresh lemon juice
2 baskets strawberries, sliced	

Bring water and sugar to a boil, stirring until sugar is dissolved. Add sliced berries and simmer, uncovered, 3 minutes. Dissolve cornstarch in lemon juice and add to the sauce, stirring over low heat until glossy and slightly thick, another 3 minutes or so. Cool. Store in refrigerator or freezer containers, allowing 1/2-inch headspace. Makes about 2-1/2 pints

FROZEN STRAWBERRY PURÉE

Serve hot over pancakes and waffles, or cool as a filling for cakes and tarts.

1 basket strawberries   1 tablespoon cornstarch
juice of 1 lemon   1/4 teaspoon salt
1 cup sugar

Coarsely crush the berries and combine them with the lemon juice, sugar, cornstarch, and salt in a saucepan. Bring to a boil over medium heat, stirring constantly. Remove from heat, cool to room temperature, and pour into freezer containers. Seal, label, and freeze.
Makes about 1 pint

STRAWBERRY SUNSHINE LEATHER

The natural fruit sugar in the berries seems sweet enough without added sugar or honey, but if you really have a sweet tooth, add some.

3 baskets strawberries

sugar, honey, or corn syrup to taste

Bring the berries slowly to a boil, stirring constantly. Purée in blender container (you should have about 4 cups). Cool to lukewarm. Meanwhile, line 2 jelly roll pans (15-1/2 by 10-1/2 inches) with clear plastic wrap, securing edges with tape. Pour lukewarm purée into pans, spreading smoothly and evenly to about 1/8-inch thickness. To keep fruit clean while drying, stretch a piece of cheesecloth over surface (but not touching purée). Place in full sunlight. Drying time depends on the fruit and the sunlight—it can vary from 8 to 24 hours. Leather is dry when it can be peeled off plastic easily. Don't leave in sun longer than necessary. Bring indoors at night and continue drying the second day, if necessary.

Drying may be finished in the oven if you run out of sun. Preheat oven to lowest temperature setting (140° to 150°). Turn oven off and set sheets of purée on center rack. If you have an electric oven you may have to reheat oven occasionally to maintain warmth. The pilot light in gas ovens is usually sufficient to dry the fruit. Open oven door every few hours to let moisture escape. The leather will dry in one or two days.

To store, roll in plastic wrap and seal tightly. Leather will keep at room temperature about 1 month, in refrigerator about 3 months, or a whole year in the freezer.

Makes 2 large rolls

STRAWBERRY FREEZER JAM

Everyone marvels at the fresh strawberry flavor of this easy and popular runnier-than-jam. So good as a sauce or eaten by the spoonful.

1 heaping basket fresh strawberries
4 cups sugar
3/4 cup water

1 1-3/4-ounce package powdered fruit pectin

Crush berries thoroughly. Measure 2 cups fruit with juices into bowl. Mix in sugar. Combine water and pectin in a saucepan. Bring to a boil and boil 1 minute exactly, stirring constantly. Pour over strawberry mixture and continue stirring 3 minutes. Ladle quickly into freezer containers, allowing 1/4-inch headspace. Seal. Let stand at room temperature 24 hours. Store in freezer. Or, if used within 2 or 3 weeks, simply refrigerate.

Makes about 2-1/2 pints

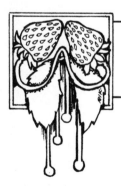

To revive delicatessen roasted chicken, rub it with melted strawberry jelly sharply seasoned with mustard and lemon juice. Pop it into a very hot oven for a few minutes to glaze, taking care that it doesn't burn.

les crèmes des crèmes

We tend to think of strawberries and cream in their simple, pristine glory. But there is a whole family of creams—pale, soft, untroubled—that rise to any occasion with strawberries.

STRAWBERRIES WITH FLORENTINE CREAM

Praline powder is elegant, easy, and keeps well in an airtight container. It is wonderful with any of these creams.

Florentine Cream
3 egg yolks
2/3 cup sugar
2/3 cup Marsala or good sherry
1 cup whipping cream, whipped

Praline
1-1/2 cups sugar
1/2 cup water
1 cup blanched almonds

2 baskets strawberries, sliced and sweetened

For the lovely, soft cream, beat egg yolks in top of double boiler until thick and pale. Gradually add sugar, beating constantly. Stir in Marsala. Place over simmering water and cook until mixture thickens, about 30 minutes, stirring occasionally. Remove from heat. Chill. Fold in whipped cream.

To make praline powder, combine sugar and water in a heavy skillet. Add almonds. Cook, stirring, until sugar dissolves and mixture becomes a light, golden caramel color. Quickly remove from heat and pour on lightly oiled baking sheet. When cool, pound or blend to desired fineness.

To serve, spoon berries into dessert glasses and top with cream and praline powder.
Makes 6 servings

STRAWBERRY CRÈME BRÛLÉE

1 quart light cream or half-and-half
1/4 cup granulated sugar
2 teaspoons vanilla extract
4 large eggs, beaten

1-1/2 cups firmly packed light brown sugar
1 basket large strawberries
powdered sugar

Heat the cream with granulated sugar and vanilla, stirring until sugar has dissolved. Add a little hot mixture to the eggs, beating constantly; then carefully stir in rest of hot mixture. Pour through a fine sieve or a piece of cheesecloth into a 1-1/2-quart mold or baking dish. Set in a pan of hot water (the water should come about halfway up the mold). Bake in a preheated 325° oven 45 minutes until a knife inserted comes out clean. Chill, covered, at least 3 hours. Sprinkle the brown sugar over the surface—a generous 1/4 inch. Slip on the lowest rack in broiler and let the sugar glaze and melt, watching all the time or it will burn. Cover and chill. Arrange whole berries on top and sprinkle lightly with powdered sugar.
Makes 6 servings

AN EASY SOUR CREAM BRÛLÉE

In this case, the berries are on the bottom of the *brûlée,* and it's all done so quickly in the broiler that the berries are scarcely heated.

1 8-ounce package cream cheese, softened
1 cup sour cream

2 baskets large berries
light brown sugar to cover surface thickly

Beat cream cheese until fluffy. Fold into sour cream. Arrange whole berries upright in a shallow baking dish. Spoon sour cream mixture on top. Sprinkle thickly with brown sugar and slip under the broiler, on the lowest rack, until the sugar melts, watching carefully to see that it doesn't burn. Chill.
Makes 6 servings

ALMOST CRÈME FRAÎCHE

Here are two ways to make the legendary *crème fraîche* of France, not exactly sour, but with a definite bite. One is richer, of course, but the one with light cream has a softer texture. These can be refrigerated for 2 or 3 weeks.

Crème Fraîche I
1 cup sour cream
4 cups whipping cream

Crème Fraîche II
1 quart light cream or half-and-half
1/4 cup buttermilk

Mix sour cream and whipping cream (unwhipped) or light cream and buttermilk in a bowl. Cover and let stand in warm place about 24 hours. It will be thick. Store in refrigerator.
Makes 1 quart

STRAWBERRIES WITH BRANDIED HONEY AND CREAM

What nicer way to use your homemade Crème Fraîche.

1/2 cup honey
1/4 cup fresh lemon juice

1/4 cup Cognac or other brandy
2 baskets strawberries, halved

Mix honey with lemon juice and Cognac. Pour over berries and chill several hours. Spoon into stemmed glasses and serve with anything creamy.
Makes 6 servings

MILK CREAM

Although the sound of sweetened condensed milk doesn't exactly turn me on, it makes some deceptively delicious sauces to spoon over berries and other fruit. The good Mexican cooks caramelize it, can and all, in boiling water for about an hour (make sure that the can is completely covered with water all the time). Here is another nice trick.

1 14-ounce can sweetened condensed milk
1/3 cup fresh lemon juice

1/2 cup whipping cream, whipped
2 baskets large strawberries with caps and stems if possible

Whip the milk in your electric mixer with the lemon juice until thick. Fold in the whipped cream and chill well. Spoon into little pots and use for dipping the berries.
Makes 6 servings

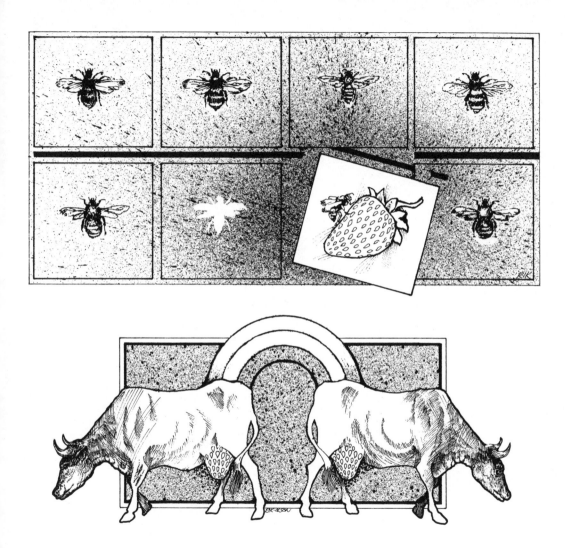

ROMANOFF, ROMANOFF, ROMANOFF

There are as many Romanoffs as there are Stroganoffs, ranging from the sublime to the not-so. A real Romanoff should be soft, fragrant, simple—any of these will do nicely.

Romanoff I

2 baskets strawberries
2/3 cup sugar
1/3 cup kirsch

2 cups whipping cream
1 teaspoon vanilla extract

Put strawberries in a bowl, sprinkle with 1/3 cup of the sugar and the kirsch. Marinate in the refrigerator an hour or so, turning gently. Whip the cream just until soft peaks form; fold in the rest of the sugar, the vanilla, and the berries in their syrup. Pile into dessert glasses and chill until serving.

Romanoff II

2 baskets strawberries
1/3 cup sugar
1/3 cup kirsch

1 cup whipping cream
1 pint soft vanilla ice cream
dash maraschino liqueur

Marinate berries in sugar and kirsch. Whip cream and fold into ice cream with a dash of maraschino liqueur. Pour over berries in stemmed glasses.

Romanoff III
2 baskets strawberries
1/3 cup Grand Marnier

1 cup whipping cream
1/3 cup sugar

Marinate berries in Grand Marnier. Purée half of them. Whip cream and sweeten with sugar; fold in purée. Spoon over whole berries in dessert dishes.
Makes 6 servings

FLORENTINE FROZEN CREAM

Any of those golden Italian liqueurs add a very special flavor to this gorgeous cream.

1 cup sugar
1/4 cup water
4 egg yolks
2 cups macaroon crumbs, moistened with liqueur

1-1/2 cups whipping cream, whipped
1/2 cup Galliano or Strega liqueur
fresh strawberries, halved figs, and clusters of grapes

Cook sugar and water, stirring until sugar is dissolved, to the soft ball stage (238° on your candy thermometer). Meanwhile use your electric mixer to beat yolks until pale and thick; then continue beating while adding syrup in a fine stream. Beat and beat until mixture is very thick and fluffy, a good 15 minutes. (This is important.) Chill until cold. Pat moistened crumbs around bottom and sides of an 8- or 9-inch springform pan. Fold whipped cream and Galliano or Strega into egg yolk mixture and pour into pan. Freeze firm. Remove from freezer 15 minutes or so before serving and top with fruit.
Makes 6 to 8 servings

FRENCH EMBASSY STRAWBERRIES

Each guest has a liqueur glass of Cognac and one of sauce, and a pile of the most beautiful long-stemmed strawberries you can find. The idea is to dip first into the Cognac, then the sauce (and, of course, sip the Cognac between bites).

Sauce
1/4 pound butter
1 cup sugar
3/4 cup whipping cream
1 egg yolk

2 tablespoons fresh lemon juice
2 tablespoons Grand Marnier
    or other liqueur (optional)

long-stemmed strawberries
plenty of Cognac

To make sauce, melt the butter in a saucepan. Stir in sugar and cream. Bring to a boil. Reduce heat and simmer 3 minutes. Stir a little of the sauce into egg yolk, then beat yolk into hot sauce. Stir 2 minutes more over low heat. Add lemon juice and liqueur. Cool.
Makes 6 servings

Strawberry Nut Butter came by way of England—a specialty at tea, and delicious on almost anything. Cream 1/2 pound sweet butter with 1 pound powdered sugar, 1/2 cup finely ground almonds and enough crushed berries to give a nice consistency for spreading.

HOMEMADE CREAM CHEESE

Perfectly delicious and with the price of cheese these days a penny pincher as well.

1 quart light cream or half-and-half
2 tablespoons buttermilk
1 teaspoon salt

mixed fresh herbs, chopped candied ginger, chopped fresh mint, grated orange or lemon peel (optional)
whole strawberries, with stems
brown sugar

Heat the cream in a saucepan until it is lukewarm (90° to 100°). Pour into a bowl and stir in buttermilk. Cover and let stand in warm place 24 to 48 hours until a soft curd forms. Line a colander with a double thickness of cheesecloth and set in sink. Pour curd into cloth and drain 15 minutes. Fold cloth over curd. Set colander in bowl large enough to allow space between bottom of colander and bowl. Cover with plastic wrap and refrigerate 12 to 18 hours. Turn curd into a mixing bowl and blend with salt and any of the seasonings listed. Mold in small straw or wire baskets or other containers. Keeps a week or more. Serve with fresh strawberries with a bowl of brown sugar for dipping.
Makes about 1 pint

Ever had a stuffed egg stuffed with a strawberry? Perfectly delicious. First, mix the sieved hard-cooked yolks with a little softened cream cheese, curry powder and chutney. Stuff eggs with this mixture and top each with a pretty berry.

MOLDED DEVONSHIRE CREAM

1 envelope unflavored gelatin
3/4 cup cold water
1 cup sour cream
1 cup whipping cream

1/2 cup sugar
1 teaspoon vanilla extract
1 basket strawberries, sliced

Soften gelatin in water. Dissolve over low heat. Cool. Stir into sour cream and set aside. Whip cream, gradually adding sugar until soft peaks form. Fold in vanilla and sour cream mixture. Pour into 1-quart mold and chill firm. Unmold on a pretty plate and ladle sliced, unsweetened berries over all.
Makes 8 servings

Note A reasonable facsimile of the real Devonshire cream can be made with whipping cream, whipped just enough to give it a little body, folded into sour cream. A cup of each. This will keep in the fridge for a couple of weeks.

COEUR À LA CRÈME

1 pint (16 ounces) creamed, small-curd cottage cheese
1 cup whipping cream, unwhipped

1 teaspoon salt
1 basket strawberries, sliced and sweetened

Whip cottage cheese as smooth as possible, gradually adding cream and salt. Line a little heart-shaped basket or dish with a layer of damp cheesecloth. Pour in cheese mixture, cover with more cheesecloth, and let stand in bowl in refrigerator overnight to drain. Next day, unmold on a serving dish and serve with sliced berries.
Makes 4 servings

STRAWBERRIES WITH ORANGE SOUR CREAM

1 basket strawberries,
   sliced and sweetened
1 cup sour cream

1 tablespoon grated orange peel
2 tablespoons fresh orange juice
dark brown sugar

Mix berries with sour cream, orange peel and orange juice. Chill for at least an hour to allow flavors to blend. Spoon into stemmed glasses and sprinkle with brown sugar.
Makes 3 to 4 servings

SIMPLE AND SUMPTUOUS SYLLABUB

The evolution of syllabubs and fruit fools has resulted in the sometimes glorious mess called the English trifle. This method brings you back to the lovely origins.

grated peel and juice of 1 lemon
1/2 cup white wine or sherry
2 tablespoons brandy
1/4 cup sugar

1 cup whipping cream
fresh nutmeg
1 basket or more strawberries, sliced
candied violets or rosemary sprigs

Put the lemon peel and juice in a bowl with the wine and brandy and let steep overnight. Next day, add sugar and stir until dissolved. Softly whip cream in a large, deep bowl with a wire whisk. Grate in a little fresh nutmeg. Slowly pour in wine mixture, continuing to whip until beautiful soft peaks form; no longer or it will become grainy. Fold in strawberries. Garnish with candied violets, a sprig of rosemary, or whatever your fancy, and serve right away.
Makes 4 servings

breakfast any time of the day

There are certain foods that defy the boundaries of time and space. Once they were "breakfast" foods. Now, nobody blinks at the time of day.

STRAWBERRIES WITH REAL MAPLE SYRUP

The strawberries should be perfect. The maple syrup should be real. That's it. Mother Nature's favorite recipe. If you can find berries with long stems, use them. Otherwise the green caps will serve the same purpose.

perfect strawberries   pure maple syrup
ice water

For each person, arrange berries on a pretty dish with a little glass of ice water with an ice cube in it and a little glass of maple syrup. Swirl the berry in the ice water, give it a shake, dip it into the syrup, bite deeply.

A FRUIT BOWL BLENDED FOR BREAKFAST

You can skip the tomatoes, if you like, but they add an intriguing something . . . .

1 basket strawberries   1/4 cup fresh lemon juice
1 papaya, peeled and seeded   1/4 cup honey
1 banana, peeled   sparkling soda
8 cherry tomatoes, stemmed   yogurt

In your blender, purée fruit with lemon juice and honey. Pour into glasses and stir in a little soda. Top with a float of yogurt.
Makes 6 servings

BAKED APPLES WITH STRAWBERRY RUM SAUCE

4 Rome Beauty or other baking apples
1/4 cup firmly packed brown sugar
1/2 cup chopped walnuts
2 tablespoons corn syrup

Strawberry Rum Sauce
1 pint soft vanilla ice cream
1 basket strawberries, sliced
rum to taste
ground cinnamon

Wash and core the apples and set in baking pan. Combine sugar, walnuts, and corn syrup and pour into apple cavities. Bake in a 350° oven for about 45 minutes or until apples are tender but not falling apart.
    Meanwhile, mix ice cream with berries and rum to taste. Spoon over warm apples and sprinkle with cinnamon.
Makes 4 servings

CHAMPAGNE SOUP

This is the only soup I've served for dessert.

6 ripe peaches
1/4 cup fresh lemon juice
1 basket strawberries

1/2 cup or more superfine sugar
1/2 cup rosé wine
1 fifth champagne

Halve and peel peaches. Set aside 6 halves and toss with lemon juice to prevent browning. In your blender, purée rest of peaches with strawberries, sugar and wine, adding more sugar, if necessary. Chill. To serve, place peach halves in glass cups or bowls. Add half the champagne to fruit purée at the table, stir gently and ladle over peaches. Pour rest of champagne into glasses and serve with the soup.
Makes 6 servings

LITTLE ONE-EGG OMELETTES

Nice and petite, a satisfying breakfast or light dessert.

For each omelette

1 egg	1 tablespoon butter
pinch of salt	1/3 cup sliced, sweetened strawberries
dash of pepper	sour cream

A 6-inch omelette pan is perfect for these. Set the pan over a hot burner. Beat the egg with salt and pepper just enough to blend it. Drop the butter into the pan and it will sizzle at once and turn light brown. Pour in the egg, swirl it around, tilting the pan, and take it off the heat. Spoon the strawberries into the center, give the pan a little shake and detach edges with a fork. Flip over the third nearest you, then the top and slip the omelette out on a warm plate. Top with sour cream—and more berries for garnish, if you have them handy.

INDIAN PUDDING WITH CRUSHED BERRIES

The most comforting of puddings, probably not too different from the ones the Indians made with wild berries.

- 5 tablespoons yellow cornmeal
- 4 cups milk, scalded
- 2 eggs, beaten
- 2 tablespoons butter
- 1 cup maple syrup
- 1 teaspoon each ground cinnamon, ground ginger, and salt
- 1 cup cold milk
- 1 basket strawberries, crushed
- 1 cup whipping cream, slightly whipped
- 1/4 cup sugar

Whisk cornmeal into scalded milk; cook and stir over low heat until thickened. Remove from heat. Add some of this mixture to the beaten eggs, then stir eggs into the pan with butter, maple syrup, and spices. Pour into a buttered 2-quart baking dish and bake in a preheated 300° oven for 45 minutes. Stir in the cold milk, blending well, and continue baking 1 hour. Crush berries coarsely and mix with cream and sugar. Pass separately to serve over warm pudding.
Makes 8 servings

APRIL IN PARIS STRAWBERRY CUSTARD

This is breakfast in bed, if you have the good sense to make it the night before. With *croissants* and steaming *café au lait*.

1 3-inch piece vanilla bean, or	3/4 cup sugar
1 teaspoon vanilla extract	4 egg yolks
1 pint light cream or half-and-half	1 basket strawberries

If using vanilla bean, cut it in half lengthwise and put it in a saucepan with the cream and sugar. Bring almost to a simmer, then turn down heat. Add a little of the hot mixture to the yolks, then stir the yolks slowly into the cream, stirring constantly until the custard thickens. Cool. Remove vanilla bean or add vanilla extract, if using. Purée berries in your blender and pour purée into little glass bowls, about halfway; fill with custard cream.

Makes 4 servings

FRESH STRAWBERRY FRITTERS

Say what you will about feathery-light batters, nothing seems to work for me as well as plain old pancake mix and water. Wonderful with Crème Fraîche or whipped cream.

1 cup buttermilk pancake mix
2/3 cup water
peanut oil

2 baskets large strawberries,
    with long stems if possible
powdered sugar

With your rotary beater, mix pancake mix with water and let stand 10 minutes. Have your oil nice and hot (475°), about 4 inches deep. Dip berries in batter, shake off excess, and fry a few at a time until golden. Drain on paper towels and dust with powdered sugar.
Makes 6 servings

AN OLD GERMAN KUCHEN

2 cups flour
1 cup sugar
2 teaspoons baking powder
1 teaspoon salt
1/4 pound butter

2 eggs, well beaten
1 cup milk
2 teaspoons ground cinnamon
1/4 cup sugar
1 basket strawberries, sliced

Sift dry ingredients together. Cut in butter with your fingertips or a pastry blender; make a well in the center. Mix eggs and milk together and pour into the well, stirring lightly with a fork to just blend ingredients. Pour into a greased 9- by 13-inch baking dish. Bake in a preheated 350° oven 30 minutes. Combine cinnamon and sugar. Spoon sliced berries over warm cake and sprinkle with cinnamon sugar. Serve warm. Makes 8 to 10 servings

STRAWBERRY AND PLUM CROQUANT

Serve warm with cream.

1 basket strawberries
6 tart plums, sliced thin
2/3 cup sugar
3 tablespoons fresh lemon juice
4 tablespoons butter

Topping
1/2 cup flour
1/3 cup sugar
dash salt
4 tablespoons butter
1/2 cup sliced unblanched almonds
1 teaspoon vanilla extract

In a 1-1/2-quart baking dish, gently toss whole berries and plum slices with sugar and lemon juice. Dot with butter. To make topping, mix flour, sugar, and salt. Cut in butter with a pastry blender; add nuts and vanilla. Sprinkle over fruit. Bake in preheated 400° oven until golden, about 15 minutes. Cool slightly.
Makes 6 servings

STRAWBERRY MORNING BREAD

Creamy cheese and bright strawberry preserves do a lot for this bread.

2 eggs
1 cup water
1 1-pound, 1-ounce package nut bread mix
1/4 teaspoon salt

2 3-ounce packages cream cheese, softened
1 tablespoon flour
1 tablespoon mixed grated orange and lemon peel
1/3 cup strawberry preserves

Beat 1 egg lightly, add water, nut bread mix, and salt, stirring until well blended. Grease and flour a 9- by 5- by 2-3/4-inch loaf pan. Pour about half the batter into the pan. Mix cream cheese with other egg, flour, and peels, and swirl in strawberry preserves. Spoon over batter and cover with remaining batter. Bake at 350° about 1 hour until loaf tests done. Let stand about 10 minutes and turn out on wire rack to cool.
Makes 1 loaf

Strawberry Scramble for a California-style breakfast. Beat 8 eggs lightly with 1/4 cup fresh-squeezed orange juice. Proceed as usual for scrambling eggs in butter, seasoning with salt, pepper, and a dash of sugar. Just before eggs are set, scramble in a cup of sliced strawberries and a few cubes of avocado.

PRALINE COOKIES AND MILK, ANYTIME

1/4 pound butter
1-1/2 cups firmly packed brown sugar
2 eggs
1-1/2 cups flour
1 teaspoon baking powder
1 teaspoon salt
2 teaspoons vanilla extract
1 cup chopped nuts

Strawberry Icing
2 tablespoons butter, melted
1/4 cup strawberry preserves
2 tablespoons light cream
   or half-and-half
sifted powdered sugar

Melt butter in saucepan. Stir in rest of cookie ingredients, mixing lightly. Spread in greased and floured baking pan (13 x 9 inches). Bake at 350° 25 to 30 minutes. Cool. To make icing, mix butter with preserves, light cream and enough powdered sugar to make desired consistency for spreading. Cut into squares or bars. Makes about 2-1/2 dozen

The South Americans like strawberries with cheese for a merienda *snack. Top toast with fresh white cheese (traditionally,* quesa fresca, *but jack or mozzarella will do). Slip under the broiler until bubbling and top with a couple of beautiful strawberry preserves.*

Strawberries blended into a breakfast nog with honey, yogurt, and an egg. Or with melon, vodka, and crushed ice for later on.

FRAGRANT WARM COFFEE CAKE

Topping
1/3 cup flour
1/2 cup slivered blanched almonds
1 teaspoon ground cinnamon
1/2 cup sugar
2 tablespoons butter

butter
1/4 cup dry macaroon crumbs

Cake
6 tablespoons soft butter
3/4 cup sugar
1 egg
1 tablespoon grated lemon peel
1 teaspoon vanilla extract
2 cups flour
2 teaspoons baking powder
1/2 cup milk

1 basket strawberries, sliced and sweetened

Mix topping ingredients coarsely with your fingertips. Set aside. Dust well-buttered 8-inch springform pan with cookie crumbs. In mixing bowl, cream butter, sugar, egg, lemon peel, and vanilla. Sift flour and baking powder together and add to creamed mixture alternately with milk. Spread batter in prepared pan; cover evenly with topping. Bake in a preheated 375° oven 1 hour or longer until toothpick comes out clean. Cool 10 minutes before removing outer ring. Slice when warm and spoon berries over each serving. Top with slightly whipped cream, if you like.
Makes 6 servings

INDIAN CORNMEAL SHORTCAKE

1/3 cup butter
2/3 cup sugar
2 eggs
1 cup buttermilk
1/2 teaspoon baking soda

1 cup yellow cornmeal
1 cup flour
1/2 teaspoon salt
Easy Strawberry Butter, following, or
sliced sweetened berries

Melt butter in large saucepan. Remove from heat and stir in sugar. Add eggs; beat until well blended. Combine buttermilk and soda. Stir into the batter with cornmeal, flour, and salt, taking care to mix until just barely blended. Pour into a greased and floured 8-inch square pan. Bake in a preheated 375° oven for 30 minutes or until golden. Cut into squares and serve warm with Easy Strawberry Butter or sliced sweetened berries.
Makes 9 servings

EASY STRAWBERRY BUTTER

1 basket strawberries
1/2 cup orange blossom honey

2 tablespoons fresh lemon juice
1/2 to 3/4 cup whipped butter

Purée berries in your blender. Pour into saucepan with honey and bring to a boil. Reduce heat and simmer 20 to 30 minutes, stirring occasionally. Add lemon juice. Cool and refrigerate. Swirl into soft butter for French toast, pancakes, and waffles.
Makes 1-3/4 cups

MY MOTHER'S HOTCAKES

Here's to Mom, a lovely lady, a lovely cook. I remember your hotcakes with nostalgia, and I remember the recipe because of all the 2's.

2 teaspoons baking soda
2 cups buttermilk
2 eggs, unbeaten
2 cups flour
2 teaspoons salt (scant)
2 tablespoons melted butter

Dissolve baking soda in buttermilk in a large bowl. Add the rest of the ingredients, stirring lightly. It is important that the batter be lumpy and streaky. Drop by the spoonful on a hot, lightly oiled griddle, turning almost as fast as little bubbles begin to break on the surface. Serve right away with hot browned butter and anything strawberry.
Makes 5 or 6 servings

BERRY BLINTZES

Pancakes
6 large eggs
2 cups water
1 teaspoon salt
2 cups flour

Filling
2 8-ounce packages cream cheese, or
1 recipe Homemade Cream Cheese, page 35
1 pint (16 ounces) small-curd cottage cheese
1/4 cup sugar
1 tablespoon each grated orange and lemon peel
dash salt

In your blender, whirl eggs, water, salt, and flour. Let batter stand 1 hour. Heat lightly buttered 7-inch skillet over medium heat. Swirl about 3 tablespoons batter into the pan to make a pancake. Fry on one side only. Continue making pancakes until batter is used up, stacking them between waxed paper as you go.

Mix filling ingredients. Spoon about 3 tablespoons filling in the center of each blintz on the browned side. Fold in ends and roll up. Chill until ready to use.

To serve, melt butter in large skillet over medium heat. Brown filled blintzes on folded side, then turn and brown other side. Serve with anything strawberry.
Makes about 2 dozen blintzes

STRAWBERRY WAFFLES WITH COCONUT SNOW

Not only any time, but any course.

2 eggs, separated
2 teaspoons baking powder
1/2 teaspoon salt
1 tablespoon sugar
1 cup milk

3 tablespoons light salad oil
1-1/3 cups flour
1 basket strawberries, sweetened
1 cup whipping cream, whipped
1/2 cup flaked coconut

Beat egg whites stiff. In another bowl, blend egg yolks with baking powder, salt, sugar, milk, and oil. Add flour gradually, stirring just until smooth. Fold in egg whites and bake in waffle iron according to manufacturer's directions. Serve warm with strawberries and top with whipped cream and coconut.
Makes 4 large waffles

sweet and sharp salads and compotes

In Venice, at the height of the Alpine strawberry season, they are served with a wedge of lemon and sugar. No cream. In France, a splash of red wine vinegar to sharpen the flavor. It seems every country has its own special way of bringing out the flavor of this popular international berry.

STRAWBERRIES IN SANGRIA SAUCE

1 10-ounce jar currant jelly
1/2 cup burgundy or other red wine
juice of 1 orange and 1 lemon

2 baskets strawberries
slices of lemon and orange

In a small saucepan, heat jelly with wine, stirring until jelly has melted. Remove from heat. Cool. Add juices. Pour over berries and chill thoroughly. Garnish each serving with a slice of lemon and a slice of orange.
Makes 6 to 8 servings

STRAWBERRIES CASSIS

In Dijon, most everybody drinks "kir"—chilled white wine with a splash of cassis, the local black currant liqueur. Nice for lunch with little sandwiches, or as a dessert. It is even better with strawberries for spooning and sipping.

2 baskets strawberries
1 bottle fruity white wine, chilled

1 bottle cassis

Spoon the berries into oversized wineglasses, pour the white wine over them, and pass the bottle of cassis.
Makes 6 servings

STRAWBERRIES WITH STRAWBERRY MAYONNAISE

I got this idea from the old Palace Hotel in San Francisco. If time is running out, you can do the same with store-bought mayonnaise.

1 egg
1 tablespoon fresh lemon juice
1-1/2 teaspoons red wine vinegar
1/2 teaspoon salt
3/4 cup light salad oil

1 tablespoon sugar
2 baskets strawberries, sliced and
   slightly sweetened
watercress or salad greens

In your blender, blend egg, lemon juice, vinegar, and salt. Put blender on medium-low speed and gradually add oil in a slow, steady stream until mayonnaise is thick. Blend in sugar and a few berries to color it softly pink. Chill. Splash sweetened berries with a little vinegar. Chill. Spoon berries over watercress or salad greens and top with a spoonful of mayonnaise.
Makes 8 servings

STRAWBERRIES IN STRAWBERRY SAUCE

This is especially nice when all you have are frozen berries.

8 cups frozen strawberries
sugar if needed

1/4 cup fresh lemon juice
2 tablespoons kirsch or
   maraschino liqueur

Purée 2 cups of the berries in your blender, sweetening with a little sugar if necessary. Add lemon juice and kirsch or maraschino liqueur. Keep the rest of the berries in the refrigerator to half-thaw. When you're ready to serve, spoon them onto your prettiest dessert plates, top with the purée, and pass the cream.
Makes 6 servings

STRAWBERRY WALDORF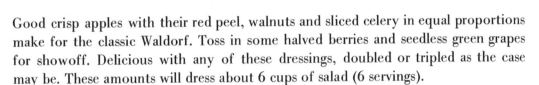

Good crisp apples with their red peel, walnuts and sliced celery in equal proportions make for the classic Waldorf. Toss in some halved berries and seedless green grapes for showoff. Delicious with any of these dressings, doubled or tripled as the case may be. These amounts will dress about 6 cups of salad (6 servings).

Dressing I Fold together 1/2 cup whipping cream, whipped, 1/2 cup mayonnaise, 1 teaspoon fresh lemon juice, 1 tablespoon sugar, and a dash of salt.

Dressing II Fold together 1 cup mayonnaise, 1/2 cup light corn syrup, and 1 cup sour cream.

Dressing III Mash the yolks of 2 hard-cooked eggs, work in 2 tablespoons fresh lemon juice and 2 cups sour cream. Add salt and pepper to taste.

A BOWLFUL OF JELLIED STRAWBERRIES

Like most of these salad recipes, this one doubles as dessert. I always think of gelatin salads as "church supper" affairs, and for some reason they're always the first to be polished off. You can make this with store-bought strawberry sherbet or Pure and Simple Strawberry Ice.

1 3-ounce package strawberry-flavored gelatin
2 cups boiling water
1 pint strawberry sherbet
1 basket strawberries
1 cup sour cream
1/2 cup whipping cream, whipped
whole berries, curly lettuce, geraniums, or other garnish of choice

In a large bowl, mix gelatin with boiling water, stirring to dissolve. Whisk in sherbet until it melts. Chill until slightly thickened. Fold in strawberries and pour into a serving bowl. Chill. Combine sour cream and whipped cream and spoon on top. To serve, border the bowl with whole berries, bright, curly lettuce and maybe some geraniums.
Makes 8 servings

A lovely dressing for fruit salad and a lovely yogurt shake. In your blender, whirl 1 cup yogurt with a basket of strawberries, a peeled papaya, a banana, and a mango. Add lemon juice and honey to taste. Chill.

THE HOLLYWOOD BOWL

One of the stars that came out of Hollywood in the golden years was "The Fruit Bowl," an extravaganza in the style of Cecil B. De Mille. Such a creation might combine sliced peaches, pineapple fingers, bing cherries, sliced persimmons, peeled figs, huge fresh desert dates, avocados, grapes, and, naturally, strawberries of incredible size and color. Here are a few ideas around which to build your own production.

Frosted Grapes and Strawberries Snip grapes into small clusters. Wash and dry large berries, preferably with long stems. Whisk a couple of egg whites until frothy; dip fruit first in egg white, then in superfine granulated sugar. Let dry on waxed paper. Use to garnish your salad plates.

Stuffed Papaya Rings Cut stem ends off papayas and scoop out seeds with a melon baller. Stuff with whipped cream cheese mixed with a little shrimp, lobster, crab, or what have you. Chill. Slice into rings and fan out in center of fruit bowl.

Pineapple Boats with Ham and Cheese Quarter a whole pineapple, cutting through crown. With a sharp knife, slice through pineapple, leaving shell intact, making 1/2-inch thick triangles. Slip triangles of smoked cheese and ham between triangles of fruit.

Melon in Port with Strawberries Cut small melons in half and scoop out seeds. Fill with sweetened sliced strawberries and a generous splash of good port. Chill before serving. Surround with greens and sliced fruit.

POLYNESIAN FRUIT BOWL WITH WHIPPED RUM

2 cups sugar
2 cups water
1/2 cup fresh lemon juice
1/2 cup dark rum

1 basket strawberries, halved
1 small pineapple, cut in chunks
6 egg yolks
3 papayas, halved and seeded

In saucepan, bring sugar and water to a boil, stirring until sugar dissolves. Simmer 15 minutes. Chill. Add lemon juice and rum. Stir berries and pineapple into marinade and chill several hours or overnight. Drain off 2/3 cup marinade from fruit bowl. Beat yolks in top of double boiler until pale. Gradually beat in the 2/3 cup rum marinade. Set over simmering water and cook, stirring occasionally, until thickened, about 30 minutes. Remove from heat and cool to room temperature. Spoon fruit into papaya halves and top with whipped rum.
Makes 6 servings

A GREEK SALAD WITH WHOLE ORANGES AND FETA

This fantasy should be served on one of those fabled Greek yachts.

zest of 1 orange
1-1/2 cups water
1-1/2 cups sugar
2 teaspoons orange flower water
   (optional)

8 sweet oranges
2 baskets strawberries
watercress
about 1 pound feta cheese

With vegetable peeler, remove just the outer skin—the zest of the orange—and sliver it. Simmer with water and sugar in a saucepan about 15 minutes, stirring now and then. Cool. Add orange flower water. Peel oranges, carefully removing all the white membrane and strings. Put whole oranges into a bowl, cover with the berries and orange syrup. Chill several hours. Serve on beds of watercress with a good chunk of feta on each plate.

Makes 8 servings

A CHINESE SOUFFLÉ SALAD

Lichees are available in cans in Chinese and specialty grocery stores. The lichee is a nut that is really a fruit; stuff them with cream cheese or almonds if you like.

3 3-ounce packages lemon-flavored gelatin
2 cups boiling water
2-1/2 cups cold water
1 cup sour cream
1 basket strawberries, halved
1 1-pound can lichee nuts, drained and halved
salad greens
candied ginger or chopped pistachios for garnish

Dissolve gelatin in 2 cups boiling water. Stir in 2-1/2 cups cold water. Pour 3/4 cup of this mixture into a 1-1/2-quart mold and chill. Add sour cream to remaining gelatin. Chill until slightly thickened. Whip with electric beater until light. Fold in berries and lichees. Pour into mold and chill firm. Unmold on greens and sprinkle with a little chopped candied ginger or pistachios.

Makes 8 servings

STRAWBERRIES ON THE HALF SHELL

This amazing cream and sugar dressing seems appropriate for this amazing salad.

2 avocados
juice of 1 lime
salad greens
1 basket strawberries, sliced

1/2 cup whipping cream, not whipped
1/4 cup white vinegar
2 tablespoons sugar

Cut avocados in half and remove pits. Douse with lime juice. Set on salad greens and fill hollows with sliced strawberries. Chill. For the dressing, mix cream with vinegar and sugar, stirring to dissolve sugar. Chill. Spoon over salads.
Makes 4 servings

BIRTHE'S DANISH RHUBARB ASPIC WITH BERRIES

Aspic
7 cups cut-up rhubarb (about 10
    long stalks)
2 cups water
3/4 cup sugar
1 envelope unflavored gelatin

Cream Cheese Sauce
2 egg yolks
1/4 cup sugar
2 3-ounce packages cream cheese
1/3 cup light cream or half-and-half
1 teaspoon vanilla extract

salad greens
1 basket strawberries, sliced and
    sweetened

Cook rhubarb with water and sugar until very soft and mushy, about 15 minutes. Strain through wet cheesecloth, squeezing out every drop of juice to make 3-1/2 to 4 cups. Discard pulp. Chill juice. In a saucepan soften gelatin in 1 cup cold juice; stir over low heat to dissolve. Mix with rest of juice and pour into a 1- to 1-1/2-quart ring mold and chill firm.

To make sauce, beat yolks with sugar until pale; gradually add cream cheese and cream. Add vanilla. Unmold aspic on salad greens, turn sweetened berries into center, spoon some of sauce over the salad, and pass the rest.
Makes 6 servings

STRAWBERRY WATERMELON

This calls for a picnic.

1-1/3 cups sugar
2/3 cup water
1 tablespoon grated lime peel
1/2 cup fresh lime juice

1/2 cup light rum
3 baskets strawberries, sliced
1 large watermelon

Combine sugar and water in a small saucepan and bring to a boil. Reduce heat and simmer 5 minutes. Add lime peel and let cool; then add lime juice and rum. Pour over strawberries and chill. Take the melon to the picnic in a bucket of ice. When you're ready to serve, slice it and serve it on paper plates with a ladleful of rummy berries.
Makes 12 or more servings

refreshingly frozen drinks and desserts

Probably the easiest and most elegant "ice capade" is to freeze a block of ice in a large bucket. Unmold it under hot water and chip away a bowl in the center to hold a big mound of spectacular strawberries with bright green caps or stems. A bowl of sour cream and one of raw, coarse brown sugar is as good as anything for dipping.

PURE AND SIMPLE STRAWBERRY ICE

At Les Frères Troisgros in Roanne, one of the grand three-star restaurants of France, a spectacular dessert is presented on a huge platter—a crown of vanilla ice cream and strawberry ice studded with both tame and wild strawberries, raspberries, pears and peaches, prunes—all coated with strawberry juice and thick cream. Accompanying this, various cakes, almond *tuiles, palmiers, brioche pralinée,* enough, enough! Here is my own pure and simple version. Do as you will with it.

3 baskets strawberries
3/4 cup sugar
1/4 cup fresh lemon juice
1/4 cup Grand Marnier or other orange-flavored liqueur

In blender, whirl berries smooth. Add sugar, lemon juice, and liqueur. Whirl again to dissolve. Pour into mold or trays and freeze firm. Let soften slightly before serving. Makes about 1 quart

Note If using frozen, sweetened berries, you'll need 2 10-ounce packages. Omit sugar in recipe.

THE QUICKEST STRAWBERRY ICE CREAM

This is a neat trick with frozen berries—whipped together in your blender just before serving.

1/2 cup whipping cream, not whipped
1 large egg

3 cups sweetened frozen strawberries

Pour the cream into the blender with the egg. Add a few of the berries. Turn on blender and, at high speed, gradually add the rest of the berries, whirling until smooth. If too soft, put into freezer for a bit.
Makes 4 servings

Note Unsweetened berries may be used in this recipe. Add 1/2 cup or more sugar to cream and egg and blend very quickly. Then add berries as directed above.

FRESH STRAWBERRY ICE CREAM

2 baskets strawberries
juice of 1/2 lemon
1/2 cup sugar
1/2 cup water

3 egg yolks
2 cups light cream or half-and-half, scalded and cooled
powdered sugar to taste

Purée the berries coarsely and mix with lemon juice. In a small pan, cook sugar and water over low heat, stirring until sugar has dissolved. Boil steadily until syrup spins a thread (230° to 234° on your candy thermometer). Beat yolks until pale, gradually beating in the syrup until mixture is cool and thick. Stir in cooled cream and strawberry purée. Taste for sweetening and add a little powdered sugar, if necessary. Freeze in a churn freezer according to manufacturer's directions or pour into freezer containers and freeze, stirring once or twice.
Makes about 1-1/2 quarts

STRAWBERRY CREAM JUBILEE

A spoonful of ice cream melted into the sauce makes this dessert a bit different.

4 tablespoons butter	2 tablespoons orange-flavored liqueur
1/4 cup sugar	2 tablespoons brandy
2 tablespoons water	1 quart vanilla ice cream
3 thin spirals of lemon peel	2 baskets beautiful whole strawberries

In a chafing dish, heat butter until bubbly. Add sugar, water, lemon peel, and liqueur. Stir over high heat until sugar dissolves and mixture is syrupy (2 or 3 minutes). Warm brandy; ignite and pour into syrup. When flame dies, stir in a heaping spoonful of the ice cream. Add berries, spooning sauce over them to coat. Immediately spoon over ice cream in stemmed glasses at the table.
Makes 6 servings

MOLDED STRAWBERRY SUNDAE

The alcohol in the liqueur keeps the strawberries saucy even when frozen.

2 baskets strawberries
1/3 cup kirsch or other fruit brandy
1 quart best vanilla ice cream
1 cup whipping cream, whipped and sweetened
1/2 cup pistachios, chopped

Set aside a few of the prettiest berries for garnish and slice and sweeten the rest. Marinate the berries an hour or two in the kirsch. Let the ice cream soften slightly and spread it about 1-inch thick in a 2-quart melon mold, saving enough to spread over the top. Fill with the berries and their juice. Cover with remaining ice cream, then waxed paper and the lid, if you have one. Freeze several hours until firm. Unmold on a serving platter. Pipe whipped cream around the edge, garnish with berries and shower with pistachios. Put back in freezer until serving time and slice before their eyes.
Makes 8 servings

A REAL STRAWBERRY PARFAIT

The French word *parfait* means "perfect," and so it is.

1/2 cup sugar
1/2 cup water
2 egg whites
1 teaspoon almond extract
1/4 teaspoon salt

1 cup whipping cream, whipped
2 baskets strawberries, sliced, sweetened, and marinated in 1/2 cup maraschino, kirsch, or other liqueur

In saucepan, boil sugar and water over medium heat to soft ball stage (238° on your candy thermometer). Meanwhile, in your largest electric mixer bowl, beat egg whites stiff, but not dry. Gradually beat in syrup. Add almond extract and salt, mixing until well blended. Fold in whipped cream. Alternate layers of cream mixture and marinated strawberries in parfait glasses. Freeze firm. Remove from freezer to thaw slightly before serving, about 20 minutes.

Makes 6 servings

STRAWBERRY CREAM WITH RASPBERRIES

2 baskets strawberries
1/4 cup kirsch
1 cup superfine sugar

1/2 cup whipping cream, not whipped
1 box raspberries

Sieve the strawberries. Add the kirsch, sugar, and cream, gently mixing to combine. Chill well. Serve in champagne glasses and top with raspberries.
Makes 4 servings

STRAWBERRY MOUSSE WITH AN ICY HEART

I once had a haunting dessert at the famed San Angel Inn in Mexico City—haunting because I tried many times to duplicate it. A pale soft cream on the outside with an icy heart, and I first tasted it with strawberries. You can simply chill this for a mousse, freeze it, or freeze it and let it half-thaw.

3 baskets strawberries
1-1/2 cups sugar
1/4 cup fresh lemon juice
1/4 cup Grand Marnier or kirsch
2 tablespoons unflavored gelatin

1/2 cup cold water
3 cups whipping cream, whipped
strawberries and whipped cream
    for garnish

Set aside half a dozen of the best berries for garnish and sieve the rest. Add sugar, lemon juice, and liqueur. Soften gelatin in water and stir over hot water to dissolve. Add to the purée and chill until slightly thickened. Fold in whipped cream. Pour into a 2-1/2-quart dish. Cover with plastic wrap and chill or freeze. Let stand at room temperature to half-thaw before serving (about 2 hours). Garnish with strawberries and more whipped cream, if you dare.
Makes 8 servings

MARY GYLLING'S MOUSSETRAP

1/4 pound butter, melted
1/4 cup firmly packed brown sugar
1 cup flour
1/2 cup chopped nuts
2 egg whites

1 cup sugar
1-1/2 baskets strawberries
2 tablespoons fresh lemon juice
1 teaspoon vanilla extract
1 cup whipping cream, whipped

With a fork, mix the melted butter, brown sugar, flour, and nuts. Spread out evenly on a rimmed cookie sheet or a jelly roll pan and bake at 400° for 15 to 20 minutes, stirring occasionally. Let cool.

In your largest electric mixer bowl, beat egg whites until almost stiff; gradually add sugar. Add berries, lemon juice, and vanilla. Beat on high speed until glossy, stiff peaks form. Fold in whipped cream. Spread half the crumbs in bottom of a 9-inch springform pan. Top with strawberry mixture, then remaining crumbs. Freeze. Remove outer ring of pan and garnish with whole berries.
Makes 12 servings

Make your own strawberry wine. Purée a basket of berries and marinate with a bottle of burgundy and sugar to taste for 48 hours. Strain. Keep in refrigerator, adding a splash of soda and a squeeze of lemon before serving over ice.

FLAMING FRAISES DES BOIS NORMANDE

Although I'm not one for fireworks at the table, this combination, famous in Normandy, flamed with Calvados or apple brandy, is out of this world. It's not likely that you'll find the little wild "strawberries of the woods," but your own berries will do nicely. This is easy to do at the table in a chafing dish.

4 tablespoons butter
1/2 cup sugar
1/2 teaspoon ground cinnamon
1 tablespoon grated lemon peel
1 tablespoon fresh lemon juice

1/2 cup Calvados (apple brandy)
1 large, firm apple, unpeeled, sliced thin
1 basket strawberries
1 quart vanilla ice cream

Melt butter; stir in sugar, cinnamon, lemon peel and juice, and 1/4 cup of the apple brandy. Simmer gently. Add the sliced apple and poach a bit in the syrup; then add berries. Ignite the rest of the warmed brandy and pour over the sauce. Ladle hot over ice cream in individual glasses.

Makes 6 servings

Now that the cocktail hour features light wines and nonalcoholic drinks, consider the strawberry for an hors d'oeuvre. With a creamy dip, skewered with a scallop, or to brighten your cheese platter.

CHAMPAGNE GLACIER

This is most dramatic in a big glass punch bowl when you're serving a bunch.

1 cup sugar
1/2 cup water
4 egg whites
1 quart lemon ice

2 quarts mixed fruit (strawberries, blueberries, grapes, peaches, pears, plums, figs, nectarines, or other seasonal fruit)
1 fifth champagne or more

In a heavy saucepan, bring sugar and water to a boil, stirring until sugar dissolves. Boil to 230° on your candy thermometer, until syrup spins a fine thread. In your mixing bowl at high speed, beat egg whites stiff; pour hot syrup over whites, beating constantly. Add ice by spoonfuls, blending well. Freeze in a large round bowl. Unmold in punch bowl and pour mixed fruit over mold. Just before serving, add champagne. Ladle ice, fruit, and champagne into oversized wineglasses.
Makes 12 servings

FROZEN STRAWBERRY WINE

Versatile, this. It has the texture of icy snow and can be used in drinks, over fruit compotes or ice cream, and it can surely stand alone. For one of the best wine coolers ever, fill wineglasses with the frozen wine mixture, pour in more of the same fruity wine, and serve with straws and beautiful berries for garnish.

1 cup sugar
1/2 cup water
grated rind and juice of 1 lemon

1 basket strawberries, coarsely crushed
1 fifth fruity wine (like Moselle, chenin blanc, a good rosé)

In saucepan, bring sugar and water to a boil, stirring until sugar dissolves. Boil 3 minutes. Cool. Add lemon rind and juice, crushed berries, and wine. Freeze. Makes about 1-1/2 quarts

SLEEPING BEAUTIES

This is a curiously refreshing version of the "martini." Make ice the day before.

Gin Ice
1-1/2 cups water
6 tablespoons gin

Vermouth Syrup
2 cups water
1 cup sugar

1 cup dry vermouth
4 lemon slices

1 quart mixed fruit (peaches, plums, grapes, pears, cherries, etc.)
1 basket strawberries

Mix the 1-1/2 cups water with the gin and pour into a shallow pan. Freeze, stirring once or twice. Meanwhile, mix the 2 cups water with sugar in a saucepan. Bring to a boil, stirring until sugar dissolves. Boil 5 minutes. Cool. Stir in the vermouth and lemon slices. Cut up your fruit and drop with strawberries into the syrup. Cover and refrigerate. To serve, spoon fruits with syrup into stemmed chilled glasses and top each with a spoonful of gin ice.
Makes 8 to 10 servings

BUBBLING STRAWBERRY LEMONADE

Your bubbles can come from soda water or champagne—depending on your taste.

1 basket strawberries
1 6-ounce can frozen pink lemonade concentrate, unthawed
ice
champagne or soda water

Put berries and lemonade concentrate in your blender and purée. Fill frosty glasses with ice and about 1/2 cup purée. Pour in champagne or soda and serve with straws.
Makes 4 servings

STRAWBERRY SANGRIA

Strawberries and burgundy have an indefinable affinity.

1 3-inch strip of orange peel
1/3 cup or more superfine sugar
1/4 cup each brandy and Cointreau
1 fifth burgundy
1 basket strawberries
1/3 cup fresh lemon juice
1 cup fresh orange juice
1 10-ounce bottle club soda, chilled
orange slices
crushed ice

In a large bowl, bruise the orange peel with the sugar to release the oils. Add brandy, Cointreau, and wine. Crush half the berries, reserving the rest for slicing, and add to wine mixture. Let steep several hours in the refrigerator. Strain into punch bowl or pitcher. Add lemon juice, orange juice, sliced berries, and soda just before serving. Garnish with orange slices and serve over crushed ice.
Makes 8 servings

FRESH STRAWBERRY VODKA

1 basket strawberries
1 6-ounce can frozen apple juice concentrate, unthawed
1 cup vodka
1/4 cup fresh lime juice
2 cups crushed ice
1/4 cup whipping cream

Set aside 6 of the nicest berries for garnish. Put remaining berries in blender with apple juice concentrate, vodka, lime juice, and crushed ice. Whirl smooth. Add cream and swirl again. Garnish each drink with a strawberry.
Makes 6 servings

A BLUSHING FRAPPÉ

You can drink your strawberries a hundred ways, every one of them good. You don't even need a blender, but it helps.

2 baskets strawberries
1/4 cup fresh lemon juice
1 cup cold milk
1/3 cup sugar
1 pint soft vanilla ice cream
1 quart crushed ice

Purée the berries with the rest of the ingredients except ice. Pour over crushed ice in frosty glasses and serve with straws.
Makes 6 servings

dazzling desserts

Always save out a few of the prettiest berries for garnish—they're what make the dazzle.

GÂTEAU DE FRAISE ET FRAMBOISE

A simple cake of concentrated strawberries and raspberries in a dramatic scarlet overcoat, showered with pistachios.

- 1 20-ounce package frozen whole unsweetened strawberries, halved
- 3 10-ounce packages frozen raspberries, slightly thawed
- 1/4 cup kirsch
- 2 dozen ladyfingers (2 3-ounce packages), split
- 2 tablespoons cornstarch
- additional kirsch
- 1/4 cup coarsely chopped pistachios
- 2 cups Crème Fraîche, page 29, whipped

Early in the day, marinate berries in kirsch 1 hour. Drain berries in a colander, reserving juice. Line bottom of a 2-quart charlotte mold with circle of plastic wrap. Touch ladyfingers lightly in juice; then fit snugly into mold, lining bottom and sides with one layer of ladyfinger halves. Make alternate layers of berries and ladyfingers, ending with ladyfingers. Cover with plastic wrap. Press down with heavy plate and a weight for at least 6 hours.

Meanwhile, make sauce with reserved juice. Strain juice, adding water if necessary to make 1-1/2 cups. In a small saucepan mix the cornstarch with a little juice, then add the rest. Cook, stirring constantly, over medium heat, until sauce is thick and clear. Taste for sweetness, and add a few tablespoons more kirsch. Cool. Unmold cake on a platter and spoon glaze over the surface and down the sides. Shower with pistachios. Pass Crème Fraîche in a pretty bowl.

Makes 8 servings

A REAL STRAWBERRY SHORTCAKE

Pound cakes, sponge cakes—even hot buttered toast make nice cushions for sugary sliced berries. But *this* is strawberry shortcake.

2 baskets strawberries, sliced
1/2 cup sugar
2 cups biscuit mix
1 tablespoon grated lemon peel

1/2 cup water
1 egg, lightly beaten
4 tablespoons soft butter
sweetened whipped cream

Mix the berries gently with 1/4 cup of the sugar and chill. Blend biscuit mix with lemon peel and 2 tablespoons of the sugar, then lightly stir in the water to make a soft dough. Turn out on a floured board and knead gently 2 or 3 times. Pat out to 3/4-inch thickness. Cut with a 2-1/2-inch cutter. Place on baking sheet, brush tops with beaten egg, and sprinkle with remaining sugar. Bake in a preheated 425° oven 8 to 10 minutes, until golden. Split while warm; slather bottom halves with butter and set on individual dessert dishes. Ladle berries over buttered biscuit half, top with other half, another ladle of berries and, finally, whipped cream.
Makes 6 servings

A BEAUTIFUL CAKE

Straight from a *pâtisserie parisienne*.

2-layer white cake, your own or a mix
1/2 cup strawberry jelly
1 basket strawberries

1-1/2 cups whipping cream, whipped
  and sweetened
1 cup crumbled macaroons

Prepare the cake according to recipe or package directions. Cool. Cut out the center of one of the layers, leaving a rim of about 1-1/2 inches. Spread the bottom, uncut layer with some of the jelly and melt the rest with a little water. Carefully put the rim of the cake over the bottom layer. Arrange whole berries neatly within the rim and glaze with melted jelly. Frost the outside and top rim of cake with sweetened whipped cream and dust lightly with crumbled macaroons.
Makes 6 servings

HONEYCOMB WHOLESOME CAKE

This is how "health food" should always taste—sensational!

4 eggs, separated
1 tablespoon fresh lemon juice
3/4 cup firmly packed brown sugar
1 cup whole wheat flour
1/2 cup sliced almonds
4 tablespoons butter

1/4 cup honey
1 tablespoon milk
2 baskets strawberries, sliced and sweetened
whipped cream or just plain thick cream

Preheat oven to 300°. Beat egg yolks until pale yellow; blend in lemon juice, sugar, and flour. Beat egg white stiff and fold into the yolk mixture. Turn into greased and floured 9-inch cake pan. Bake for 40 to 45 minutes, until cake tests done. Invert on cake rack and cool for 1 hour or more. Turn upright on cookie sheet.

    Combine nuts, butter, honey, and milk in a saucepan. Cook over low heat, stirring constantly, until bubbly. Spread on cake and slip under broiler for a few minutes, watching carefully, until mixture bubbles. Cool and let glaze harden. Cut into wedges and top with strawberries and cream.
Makes 6 servings

AN AFTER-DINNER COFFEE CAKE

Wait until you taste this cake.

Topping
1/4 cup sugar
1/2 cup chopped pecans
1 teaspoon ground cinnamon

Cake
3/4 cup soft butter
1-1/2 cups sugar
2 eggs
1 teaspoon vanilla extract

1-1/4 cups flour
2 teaspoons baking powder
1/2 teaspoon baking soda
1/2 teaspoon salt
1 cup sour cream

Filling
1 basket strawberries, sliced and sweetened
1 cup whipping cream, whipped

In a small bowl, mix sugar, nuts, and cinnamon. Set aside. In mixer bowl, cream butter and sugar; add eggs and vanilla, beating until fluffy. Sift flour, baking powder, soda, and salt. Add sifted dry ingredients to creamed mixture alternately with sour cream. Spread half the batter in a greased and floured 9-inch tube pan. Sprinkle with half the topping mixture. Spoon on remaining batter, then rest of topping. Bake at 350° for 50 to 55 minutes, until cake tests done. Turn out of pan on a pretty platter. Fold together berries and whipped cream; pour into center of cake.
Makes 8 to 10 servings

FAMOUS NEW YORK CHEESECAKE

It makes other cheesecakes look undernourished. As I remember it, the crust was a butter-pastry affair which was difficult to execute in a springform pan. I personally prefer zwieback crumbs mixed with butter and sugar.

Zwieback Crust
1 6-ounce package zwieback
1/2 cup powdered sugar
1/4 pound butter, melted

Filling
2-1/2 pounds cream cheese, softened
1/4 cup whipping cream
1-3/4 cups sugar
3 tablespoons flour
2 teaspoons each grated lemon and orange peel
1 teaspoon vanilla extract
6 eggs

Topping
1/2 cup melted currant or strawberry jelly
2 baskets large, beautiful berries

Mix the crust ingredients together and pat in a buttered 9-inch springform pan which is 3 inches deep. In a large mixing bowl, beat cream cheese with cream, sugar, flour, lemon and orange peel, and vanilla. On high speed, beat until fluffy. Add eggs, one at a time, beating smooth. Pour into crumb-lined pan. Bake in a preheated 500° oven for 10 minutes. Lower heat to 250° and bake 1 hour more. Cool in pan on rack for 2 hours, then chill several hours or overnight.

Before serving, melt jelly and brush surface of cake with a thin layer. Arrange berries upright on cake and spoon rest of glaze over all the berries. Chill.
Makes 16 servings

RHUBARB AND STRAWBERRY PIE

A true fresh-as-springtime pie.

Pie Pastry, following
1 pound rhubarb
1 tablespoon cornstarch
1 cup sugar

1/2 cup currant jelly
2 baskets large strawberries
sweetened whipped cream

Prepare 10-inch baked pie shell as directed in recipe. Set aside to cool. Cut rhubarb into 2-inch pieces into a baking dish. Mix cornstarch with sugar and toss it with rhubarb; cover and bake at 375° for 20 minutes, or until fruit is soft. Drain off juice and set aside. Melt jelly and paint inside of pie shell. Add rhubarb juice to rest of jelly in a small saucepan and cook over medium-high heat until syrup is thick. Cool. Pour rhubarb into pie shell and arrange berries upright over entire surface. Spoon the syrup over the berries to glaze. Pass a bowl of whipped, sweetened cream.
Makes 8 servings

PIE PASTRY

1/2 cup shortening or lard
1-3/4 cups flour

1/2 teaspoon salt
3 tablespoons ice water

Cut shortening into flour and salt with a pastry blender until it is like cornmeal. Sprinkle water over the flour, mixing lightly with a fork. Gently work the dough into a ball and chill slightly. Set oven at 400°. Roll out pastry and fit into pie plate, cut edges and crimp. Line the unbaked pie shell with foil and fill with dry beans, pushing them up against the sides so they won't collapse. Bake 20 minutes. Remove from oven and lift out foil with beans. Return to oven and bake another 7 or 8 minutes until golden. Cool.
Makes 1 9- or 10-inch baked pie shell

QUEEN OF TARTS

Pie Pastry, preceding
1 8-ounce can almond paste
1 egg yolk
4 tablespoons butter

1 to 1-1/2 baskets large strawberries
1/2 cup strawberry jelly
powdered sugar

Prepare pastry dough as directed in recipe. Fit dough into individual tart shells or a large tart pan and bake as directed in recipe. Cool. Mix almond paste with egg yolk and butter. Spread over the bottom of the tarts. Arrange berries close together, upright on paste. Melt jelly with a little water; pour over berries to glaze well. Refrigerate and sieve a little powdered sugar over the berries.
Makes 6 servings

A SPECTACULAR STRAWBERRY TART
(Practically Thrown Together)

Who would ever guess that this flaky puff-paste wonder is nothing more than a package of frozen patty shells?

1 10-ounce package frozen patty shells, thawed
1 cup whipping cream, whipped
1 pint sour cream
1/4 cup or more Grand Marnier or other liqueur
2 baskets large, beautiful berries
melted strawberry or currant jelly

Squeeze together patty shells and roll out to fit a 12- to 14-inch pizza pan. Refreeze the pastry. Preheat the oven to 450° and set the pastry in the oven. Reduce temperature to 400°. Bake 15 minutes until golden and puffy. Cool. Mix whipped cream with sour cream and fold in the liqueur. Spread over pastry. Top with beautiful upright berries and glaze with jelly. Chill or serve at once.
Makes 8 to 10 servings

MISS MELENDY'S LEMONCURDIES

Lemony luscious, and that is the plain truth. You can make a spectacular tart with a field of strawberries or individual squares, each topped with a perfect berry.

2 cups flour
1/2 cup sifted powdered sugar
1/2 pound butter
4 eggs
2 cups sugar
dash salt
1/3 cup fresh lemon juice
2 baskets beautiful strawberries
powdered sugar

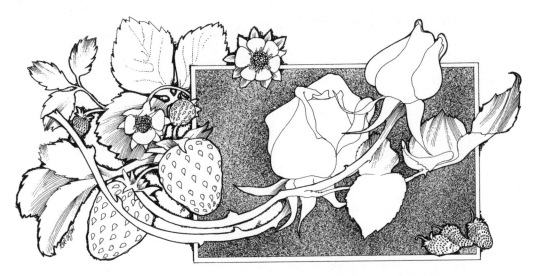

Combine flour and powdered sugar; cut in butter with your fingertips or a pastry blender. Press into a 12-inch tart pan or a 13- by 9-inch baking pan. Bake in a preheated 350° oven 20 to 25 minutes until golden.

Meanwhile, beat eggs at high speed until pale and yellow, gradually adding sugar and salt; then add lemon juice, continuing to beat at high speed. Pour over hot crust and return to oven 20 to 25 minutes longer, or until golden. Cool. Arrange berries upright on tart and sprinkle with powdered sugar. Or cut pastry into squares and top each with a berry. Dust with powdered sugar.
Makes 8 servings in a tart or 16 squares

FRESH STRAWBERRY YOGURT PIE

I discovered this in one of those "health food" places—healthy or not, who cares when it tastes like this.

1-1/2 cups crunchy whole-grain cereal (honey-sweetened granola)
1/4 pound butter, melted
1 envelope unflavored gelatin
1/2 cup cold water
2 egg yolks
1 tablespoon each lemon juice and peel

1/3 cup sugar
1 8-ounce package cream cheese
1 cup plain yogurt
4 ice cubes
1 basket strawberries, sliced
melted strawberry or currant jelly

Butter a 9-inch pie plate. Crush the crunchy cereal with a rolling pin and mix with the melted butter. Pat into the pie plate. Dissolve gelatin in water and stir over low heat until dissolved. Set aside. In your blender, blend egg yolks, lemon juice and peel, sugar, cream cheese, and yogurt. Add dissolved gelatin and blend. Add ice cubes, one at a time, blending smooth. Pour over crust and top with berries. Glaze with melted jelly. Chill.
Makes 8 servings

Strawberries and port are naturals. Plain port to sip with plain strawberries, port jelly with sliced berries, berries marinated in port.

OLD-FASHIONED CREAM PUFFS

For Gene—who asked whatever happened to . . . .

1 cup water
1/4 pound butter
1 cup sifted flour
dash each of salt and sugar
4 large eggs

2 baskets strawberries, sliced
2 cups whipping cream, whipped and sweetened
powdered sugar or Fresh Strawberry Glaze, following

Bring water and butter to a boil. Add flour, salt, and sugar all at once and beat until mixture leaves sides of pan. Remove from heat and beat in eggs, one at a time, beating well after each addition until dough is nice and shiny. Drop dough by tablespoons on an ungreased cookie sheet (or use a pastry bag), 2 inches apart to make 12 puffs. Bake 35 to 45 minutes in a preheated 400° oven until golden brown. Split while hot to keep puffs crisp. Fill with strawberries folded into whipped cream. Dust with powdered sugar or dribble with glaze.
Makes 12 cream puffs

Fresh Strawberry Glaze Beat 2 egg yolks until pale. Stir in 1/4 cup melted butter and 1/2 cup crushed strawberries. Add enough powdered sugar to make a thin glaze.

A 25-calorie Slush. Use proportions of 1/3 fruit to 2/3 crushed ice in your blender. Whirl until slushy. Pour 2-1/2 ounces into each glass and stir with a little lemon-lime diet soda.

CHINESE CRACKLING FRUIT

This is a little tricky, but once you get the hang of it, you'll have fun.

1/2 cup light corn syrup
3/4 cup water
2 cups sugar

assorted fruits: whole strawberries and seedless grapes; sliced peaches, plums, pears, nectarines, papayas, and bananas; orange sections

Combine corn syrup, water, and sugar in a 2-1/2-quart saucepan. Cook over high heat, stirring until sugar has dissolved. Boil until temperature reaches 300° on a candy thermometer. Pour syrup quickly into a chafing dish over a simmering water bath. Keep water at a gentle simmer. Your friends spear berries and other fruits with bamboo skewers, dip into syrup, and then quickly dip into individual bowls of ice water to harden syrup.
Makes 6 servings

THE RICE PUDDING OF MARTINIQUE

Soft as a Caribbean cloud, this rummy rice pudding. Serve it *their* way with a big platter of cut-up fruit and lots of lime wedges. (Incidentally, have you ever had plain vanilla ice cream with a wedge of lime for squeezing over? Delicious!)

1 quart milk
1/2 cup sugar
1 teaspoon salt
3/4 cup long-grain rice (uncooked)
2 teaspoons vanilla extract

as much rum as you like
1 cup whipping cream, whipped
assorted fresh fruits: strawberries, pineapple, papaya, mango, grapes, avocados, peaches, plums

In heavy saucepan with tight lid, bring milk, sugar, salt, and rice to a boil, stirring all the while. Reduce heat and cover. Simmer 25 to 30 minutes, stirring occasionally, until rice is tender and milk is almost absorbed. Stir in vanilla and rum. Cool to room temperature. Fold in whipped cream. Spoon into a pretty bowl and chill. Serve with fruit arranged decoratively on a platter.
Makes 8 servings

A GILDED LILY BREAD PUDDING

Save out some of the custard in the pudding to make a sumptuous strawberry sauce.

3 slices raisin bread, trimmed of crusts
3 tablespoons butter
1/4 cup firmly packed brown sugar
6 eggs plus 3 yolks, lightly beaten

1-1/4 cups sugar
2 teaspoons vanilla extract
4-1/2 cups milk, heated
1 basket strawberries, sliced

Butter a 2-quart casserole. Butter the bread and sprinkle with brown sugar. Cut bread into quarters. Combine eggs, sugar, and vanilla; slowly add hot milk. Measure 1-1/4 cups of this custard mixture into top of a double boiler and set aside. Pour remaining custard into casserole and top with bread squares. Bake in a pan of hot water in preheated 400° oven 20 to 25 minutes, until knife inserted comes out clean. Cool at room temperature while you make the custard sauce.

To make sauce, stir reserved custard mixture over simmering water until it coats a metal spoon. Cool. Fold berries into sauce. Pass sauce separately.
Makes 8 servings

CHOCOLATE, VANILLA, AND STRAWBERRIES

Everyone's favorite flavors—a rich chocolate mousse with a blanket of cream and glazed strawberries.

3 eggs, separated
3/4 cup granulated sugar
3 tablespoons orange-flavored liqueur
1 teaspoon vanilla extract
5 1-ounce squares unsweetened chocolate, broken into small pieces
4 tablespoons butter
2 cups whipping cream, whipped
1 basket large strawberries
melted strawberry jelly

Beat egg yolks, sugar, orange-flavored liqueur, and vanilla in the top of a double boiler. Cook over simmering water, stirring until slightly thickened. Add chocolate and butter, stirring until melted and smooth. Remove from heat and cool. Beat egg whites stiff and fold into chocolate mixture. Fold in half the whipped cream. Pour into a pretty glass bowl and frost with remaining cream. Chill several hours. Before serving, stand strawberries upright in cream and dribble with melted jelly.
Makes 8 servings

STUFFED STRAWBERRIES ON ICE

There will come a time when you'll find just the right strawberries for this. Snap them up.

about 30 very large strawberries	1 cup whipping cream, whipped
2 egg yolks	1/4 cup powdered sugar
2 tablespoons granulated sugar	additional powdered sugar
2 tablespoons Marsala, sherry, or port	crushed ice

From the point, slit each berry into quarters without cutting through the bottom. Chill. Beat egg yolks, granulated sugar, and wine over boiling water until mixture forms soft peaks, about 5 minutes. Immediately set in bowl of ice and continue beating until mixture is cool. Let stand in refrigerator 30 minutes. Stir together the whipped cream, powdered sugar, and the chilled mixture. Beat stiff. Fill each berry with cream, using pastry bag and decorating tip to bring cream to a pretty peak. Sprinkle berries with powdered sugar and arrange on mounds of crushed ice in stemmed glasses.
Makes 6 servings

GRACIOUS GRECIAN WAYS WITH STRAWBERRIES

The Greeks serve strawberries, half-dipped in cheese fondant, on polished leaves.

1 basket large berries with stems
1 3-ounce package cream cheese
1 egg yolk
1/4 teaspoon almond extract

1 to 2 cups powdered sugar
light cream or half-and-half
finely chopped almonds, about 1/3 cup

Select large strawberries with stems, if possible. Wash carefully and let them dry. In mixing bowl, beat cream cheese until soft and fluffy. Add egg yolk, almond extract, and powdered sugar. Add enough cream to make just the right consistency for dipping. Swirl berry in fondant and dust lightly with almonds. Let stand until firm. Serve on large green leaves polished with a little milk.
Makes 4 servings

A RICH, NUTTY GREEK CONFECTION

Reminiscent of those honey-rich pastries the Greeks are so fond of, this confection should be served in small pots with a mound of strawberries on the side.

1 1-pound jar (about 2 cups) orange blossom honey
1 cup whipping cream, whipped

1/3 cup each filberts, walnuts, and Brazil nuts, coarsely chopped
2 baskets large whole berries, with caps

Whip the honey on your electric mixer until foamy. Chill. When ready to serve, gently fold whipped cream and nuts into honey. Serve with spoons.
Makes 6 servings

BOB KAVET'S SENSATIONAL STRAWBERRIES

Charming Bob Kavet of the Strawberry Advisory Board has a particularly intriguing way with the whopping big berries you sometimes come upon. He has a special cook's hypodermic needle reserved for injecting liqueur into the hollow hearts, then he swirls them in fondant and dipping chocolate. These with a pot of espresso are *la crème de la crème*.

2 baskets very large strawberries, with caps and stems
Grand Marnier or other fruit liqueur
1 tablespoon instant coffee
1-1/2 tablespoons boiling water
2 tablespoons butter
1 egg yolk
1-3/4 cups or more powdered sugar
1 6-ounce package semisweet chocolate bits

The hollow hearts of the berries are dead center. Fill your hypodermic needle with liqueur and pierce the berry from the side, plunging about half a thimbleful into the heart. Set aside.

Meanwhile, prepare a fondant by mixing coffee with boiling water to dissolve. On high speed on your mixer, beat coffee, butter, and egg yolk, gradually adding powdered sugar until you have a smooth, shiny consistency. Swirl the bottom halves of the berries into the coffee fondant and set on a rack to dry.

When fondant is set, pour chocolate bits into top of double boiler and melt over about 1 inch of simmering water, stirring until smooth. Remove from heat and swirl berries in chocolate, leaving a ring of fondant visible. Set on rack and let harden. These must be served shortly after you make them.
Makes 4 servings

index to recipes

Apples, Baked, with Strawberry Rum Sauce, 40
Aspic with Berries, Birthe's Danish Rhubarb, 60

Birthe's Danish Rhubarb Aspic with Berries, 60
Blintzes, Berry, 50
Bob Kavet's Sensational Strawberries, 93
Bread Pudding, A Gilded Lily, 89
Bread, Strawberry Morning, 46
Butter, Easy Strawberry, 49

Cakes and Coffee Cakes
   An After-Dinner Coffee Cake, 80
   A Beautiful Cake, 78
   Fabulous New York Cheesecake, 81
   Fragrant Warm Coffee Cake, 48
   Gâteau de Fraise et Framboise, 77
   Honeycomb Wholesome Cake, 79
   Indian Cornmeal Shortcake, 49
   An Old German Kuchen, 44
   A Real Strawberry Shortcake, 78
   Strawberry Morning Bread, 46
Champagne Glacier, 72
Champagne Soup, 40
Chinese Crackling Fruit, 88
Chinese Soufflé Salad, A, 59
Chocolate, Vanilla, and Strawberries, 90
Coeur à la Crème, 36
Cookies and Milk, Praline, 47

County Fair Strawberry Preserves, 16
County Fair Strawberry Preserves, Quick, 17
Cream Cheese, Homemade, 35
Cream Puffs, Old-Fashioned, 87
Crème Brûlée, Strawberry, 28
Crème Fraîche, Almost, 29
Croquant, Strawberry and Plum, 45
Custards, see Puddings and Custards

Devonshire Cream, Molded, 36
Drinks
   A Blushing Frappé, 75
   Bubbling Strawberry Lemonade, 74
   Champagne Glacier, 72
   Fresh Strawberry Vodka, 75
   Frozen Strawberry Wine, 72
   A Fruit Bowl Blended for Breakfast, 39
   Sleeping Beauties, 73
   Strawberry Sangria, 74

Easy Strawberry Butter, 49

Florentine Cream, Strawberries with, 27
Florentine Frozen Cream, 33
Frappé, A Blushing, 75
Freezing Strawberries, 22
French Embassy Strawberries, 34
Fritters, Fresh Strawberry, 44
Frozen Strawberry Purée, 23
Frozen Strawberry Wine, 72
Fruit Bowl Blended for Breakfast, A, 39

Fruit Bowl with Whipped Rum, Polynesian, 58
Fruit, Chinese Crackling, 88

Gâteau de Fraise et Framboise, 77
Grecian Ways with Strawberries, Gracious, 92
Greek Confection, A Rich, Nutty, 92
Greek Salad with Whole Oranges and Feta, A, 58

Helen McCully's Famous Strawberry Syrup, 19
Hollywood Bowl, The, 57
Honeycomb Wholesome Cake, 79
How to Keep Your Fruit Icy-Crisp, 21

Ice Creams and Ices
   Florentine Frozen Cream, 33
   Fresh Strawberry Ice Cream, 64
   Molded Strawberry Sundae, 66
   The Quickest Strawberry Ice Cream, 64
   Pure and Simple Strawberry Ice, 63
   Indian Cornmeal Shortcake, 49
   Indian Pudding with Crushed Berries, 42

Jacques Pepin's Fraises au Soleil (Strawberries in the Sun), 15
Jams and Jellies, see Preserves
Jellied Strawberries, A Bowlful of, 56

Kuchen, An Old German, 44

Leather, Strawberry Sunshine, 24
Lemonade, Bubbling Strawberry, 74
Lemonade Jelly, Strawberry, 18
Lemoncurdies, Miss Melendy's, 84

Mary Gylling's Moussetrap, 70
Milk Cream, 30
Miss Melendy's Lemoncurdies, 84
Molds
    Birthe's Danish Rhubarb Aspic with Berries, 60
    A Chinese Soufflé Salad, 59
    Coeur à la Crème, 36
    Homemade Cream Cheese, 35
    Molded Devonshire Cream, 36
    Molded Strawberry Sundae, 66
Moussetrap, Mary Gylling's, 70
Mousse with an Icy Heart, Strawberry, 69

Omelettes, Little One-Egg, 41
Orange Sour Cream, Strawberries with, 37

Pancakes
    Berry Blintzes, 50
    My Mother's Hotcakes, 50
Parfait, A Real Strawberry, 67
Perpetual Pot of Fruit with Rum or Brandy, A, 20
Pie Pastry, 82
Pies and Tarts
    Fresh Strawberry Yogurt Pie, 86
    Miss Melendy's Lemoncurdies, 84
    Queen of Tarts, 83
    Rhubarb and Strawberry Pie, 82
    A Spectacular Strawberry Tart, 84
Praline Cookies and Milk, 47
Preserves
    County Fair Strawberry Preserves, 16
    Jacques Pepin's Fraises au Soleil (Strawberries in the Sun), 15
    Quick County Fair Strawberry Preserves, 17
    Rose Crystal Strawberry Jelly, 18
    Strawberry and Rhubarb Preserves, 17
    Strawberry Freezer Jam, 25
    Strawberry Lemonade Jelly, 18
Puddings and Custards
    April in Paris Strawberry Custard, 43
    A Gilded Lily Bread Pudding, 89
    Indian Pudding with Crushed Berries, 42
    The Rice Pudding of Martinique, 88
    Simple and Sumptuous Syllabub, 37
    Strawberry Crème Brûlée, 28
Purée, Frozen Strawberry, 23

Queen of Tarts, 83

Rhubarb and Strawberry Pie, 82
Rhubarb Preserves, Strawberry and, 17
Rice Pudding of Martinique, The, 88
Romanoff, Romanoff, Romanoff, 32
Rose Crystal Strawberry Jelly, 18

Sangria, Strawberry, 74
Sauces and Toppings
    Almost Crème Fraîche, 29
    Bright Strawberry Sauce, 23
    Embassy, 34
    Flaming Fraise des Bois Normande, 71
    Florentine Cream, 27
    Milk Cream, 30
    Romanoff, 32
    Sangria Sauce, 53
    Strawberry Cream Jubilee, 65
    Strawberry Mayonnaise, 54
    Strawberry Rum Sauce, 40
    Strawberry Sauce, 54
    Syrup, Helen McCully's Famous Strawberry, 19

Sensational Strawberries, Bob Kavet's, 93
Sherbets, see Ice Creams and Ices
Shortcake, A Real Strawberry, 78
Shortcake, Indian Cornmeal, 49
Soup, Champagne, 40
Sour Cream Brûlée, Easy, 28
Strawberries Cassis, 53
Strawberries in Sangria Sauce, 53
Strawberries in Strawberry Sauce, 54
Strawberries in the Sun, 15
Strawberries on the Half Shell, 60
Strawberries with Brandied Honey and Cream, 30
Strawberries with Florentine Cream, 27
Strawberries with Real Maple Syrup, 39
Strawberries with Strawberry Mayonnaise, 54
Strawberry and Plum Croquant, 45
Strawberry Butter, Easy, 49
Strawberry Cream with Raspberries, 68
Strawberry Mousse with an Icy Heart, 69
Strawberry Sunshine Leather, 24
Strawberry Waldorf, 55
Strawberry Watermelon, 61
Stuffed Strawberries on Ice, 91
Syllabub, Simple and Sumptuous, 37
Syrup, Helen McCully's Famous Strawberry, 19

Tarts, see Pies and Tarts
Toppings, see Sauces and Toppings

Waffles with Coconut Snow, Strawberry, 51
Waldorf, Strawberry, 55
Watermelon, Strawberry, 61

Yogurt Pie, Fresh Strawberry, 86

MAGGIE WALDRON

Maggie Waldron's special touch is very much in evidence. She is the blithe and talented spirit behind some of the country's most imaginative food and wine photographs. As Creative Director for Botsford Ketchum, a firm specializing in our national fresh agricultural bounty, her work appears regularly in magazines, newspapers and on film.

She is a born cook who grew up in the restaurant/hotel business, graduated from Cornell, was an associate food editor at *McCall's,* a television director for a major food company and the principal in her own consulting firm. She has recently returned from Paris with her *diplôme* from France's newest cooking school, La Varenne. The recipes in this book were developed in Maggie's Kitchen, a new food center in San Francisco.

RIK OLSON

An artist versatile in many media, Rik Olson received his BFA degree from California College of Arts and Crafts and later spent eight years in Europe as an arts and crafts instructor for the United States Army. While he was abroad, his graphics and photographs were widely exhibited in Germany and Italy, winning a number of awards. In addition to the Edible Garden Series, Rik Olson also illustrated another book, *One Pot Meals,* for 101 Productions. He and his wife presently live in San Francisco.